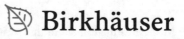

Annals of the International Society of Dynamic Games

Volume 16

More information about this series at http://www.springer.com/series/4919

David Yeung • Shravan Luckraz • Chee Kian Leong
Editors

Frontiers in Games and Dynamic Games

Theory, Applications, and Numerical Methods

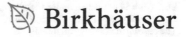 Birkhäuser

Editors
David Yeung
SRS Consortium for Advanced Study
Hong Kong Shue Yan University
Hong Kong, Hong Kong

Center of Game Theory
Saint Petersburg State University
St. Petersburg, Russia

Chee Kian Leong
School of Economics
University of Nottingham Ningbo China
Ningbo, China

Shravan Luckraz
School of Public Finance and Taxation
Zhejiang University of Finance
and Economics
Hangzhou, China

ISSN 2474-0179   ISSN 2474-0187   (electronic)
Annals of the International Society of Dynamic Games
ISBN 978-3-030-39791-3   ISBN 978-3-030-39789-0   (eBook)
https://doi.org/10.1007/978-3-030-39789-0

Mathematics Subject Classification: 91A05, 91A06, 91A10, 91A15, 91A23, 91A25, 91A28, 91A65, 91A80, 49N70, 49N90, 49N30

This book is published under the imprint Birkhäuser, www.birkhauser-science.com by the registered company Springer Nature Switzerland AG.
The registered company address is: Gewerbestrasse 11, 6330 Cham, Switzerland

Preface

The ISDG-China Chapter Conference on Dynamic Games and Game Theoretic Analysis was held from August 3 to 5, 2017, at the Ningbo (China) campus of the University of Nottingham. The plenary lectures of this conference were given by David Yeung, Distinguished Research Professor at the Hong Kong Shue Yan University; Leon Petrosjan, Professor of Applied Mathematics and the Head of the Department of Mathematical Game theory and Statistical Decision Theory at the St. Petersburg University, Russia; Georges Zaccour, Professor of Decision Sciences and Chair in Game Theory and Management at HEC Montréal in Canada; and Shmuel Zamir, Professor at the Center for the Study of Rationality, The Hebrew University in Israel.

Following the conference, a call for papers for this volume was announced and participants of the conference were encouraged to submit papers for consideration. Each paper was reviewed by at least two experts in game theory, chosen by the editors. Revisions based on the comments of the reviewers were sought for papers and acceptance was contingent on papers having revised adequately to meet the standards required by the reviewers and editors.

This volume is composed of eight accepted papers covering a variety of topics from theory to applications of games and dynamic games. All contributed papers submitted for consideration of publication to the Annals were peer reviewed according to standards of international journal and ISDG Annals.

The first two papers in this volume are based on the plenary lectures and share the common theme of cooperative dynamic games. They are authored by the two pioneers and foremost authorities in this field. Both are based on their plenary lectures during the conference.

In the paper "Dynamically Stable Cooperative Provision of Public Goods Under Non-transferable Utility", based on his plenary lecture, David Yeung considers the cooperative optimization required to deal with the classic case of market failure, namely the provision of public goods. Cooperation cannot be dynamically stable unless the participating agents' cooperative payoffs are guaranteed to be no less than their non-cooperative payoffs throughout the cooperation duration. His paper derives dynamically stable cooperative solutions for public goods provision by asymmetric agents with non-transferable utility/payoffs in a differential game framework. The paper examines the dynamically stable subcore under a constant payoff weight. In the absence of a dynamically stable subcore under a constant weight, a variable payoff weights scheme leading to a dynamically stable cooperative solution has to be designed. This paper's key innovation is the provision of a non-transferable payoff framework for ensuring dynamically stable cooperative provision of public goods.

Along the same line of cooperative dynamic games, the paper "Strongly Time-Consistent Solutions in Cooperative Dynamic Games" by Leon Petrosyan investigates the evolution of dynamic game along the cooperative trajectory. Along cooperative trajectory at each time instant players find themselves in a new game which is a subgame of the originally defined game. In many cases the optimal solution of the initial game restricted to the subgame along cooperative trajectory fails to be optimal in the subgame. To overcome this difficulty, he and his coauthors have introduced the special payment mechanism imputation distribution procedure (IDP), or payment distribution procedure (PDP), but another serious question arises: under what conditions the initial optimal solution converted to any optimal solution in the subgame will remain optimal in the whole game. This condition we call strongly time-consistency condition of the optimal solution. If this condition is not satisfied players in reality may switch in some time instant from the previously selected optimal solution to any optimal solution in the subgame, and as a result realize the solution which will be not optimal in the whole game. This paper proposes different types of strongly time-consistent solutions for multicriterial control, cooperative differential and cooperative dynamic games.

The next two papers consider games with hierarchical structure or Stackelberg games.

In his exhaustive survey entitled "Incentive Stackelberg Games for Stochastic Systems," Hiroaki Mukaidani considers dynamic stochastic incentive Stackelberg games for deterministic and stochastic linear systems with external disturbance. Although the incentive Stackelberg strategy has been admitted as the hierarchical strategy that induces the behavior of the decision maker as that of the follower, the followers optimize their costs under incentives without a specific information. Therefore, leaders succeed in using the required strategy to induce the behavior of their followers. This concept is considered very useful and reliable in some practical cases. The induced features of the hierarchical strategy in the considered models, including stochastic systems governed by Ito stochastic differential equation, Markov jump linear systems, and linear parameter varying (LPV) systems, are explained in detail. Furthermore, basic concepts based on the H_2/H_∞ control setting for the incentive Stackelberg games are reviewed. Next, it is shown that the required set of strategies can be designed by solving higher-order cross-coupled algebraic Riccati-type equations. Finally, to simulate future research in this areas, some open problems are discussed.

On the other hand, Olga I. Gorbaneva and Gennady A. Ougolnitsky in their paper "Social and Private Interests Coordination Engines in Resource Allocation: System Compatibility, Corruption, and Regional Development" apply static Stackelberg and inverse Stackelberg games to model administrative and economic control mechanisms providing system compatibility in the static game-theoretic models of resource allocation between social and private activities. Descriptive and normative approaches to the modeling of corruption in resource allocation in the hierarchical control systems are proposed and implemented, and applications to the problems of regional development are outlined.

Games on graphs and networks constitute new frontiers in the field of game theory, and the next two papers contribute to this new frontier.

Vasily V. Gusev's paper "A Multi-Stage Model of Searching for Two Mobile Objects on a Graph" deals with a search game where one searcher looks for two mobile objects on a graph. The searcher distributes his searching resource so as to maximize the probability of detecting at least one of the mobile objects. Each mobile object minimizes its own probability of being found. In this problem, the Nash equilibrium is defined as the optimal transition probabilities of the mobile objects and the optimal values of the searcher's resource. Besides establishing the Nash equilibrium, the paper also obtains the value of the game in a single-stage search game with non-exponential payoff functions.

The paper "The Impact of Product Differentiation on Symmetric R&D Networks" by Mohamad Alghamdi and his coauthors examines the impact of product differentiation on R&D networks. They find that when firms produce goods that are complements or independent, R&D expenditure, prices, firms' net profits, and total welfare are always higher under price competition than under quantity competition. When goods are substitutes, R&D expenditure and profits are higher under quantity competition than under price competition. Also, when goods are substitutes and products are sufficiently differentiated, then total welfare is higher in the Bertrand equilibrium than under the Cournot equilibrium. Beyond this threshold level of product differentiation, Cournot competition is superior in terms of social welfare. The key threshold level of product differentiation, determining the relative superiority of the Cournot and Bertrand equilibrium when goods are substitutes, depends on the cost efficiency of R&D and the number of collaborative partnerships that firms participate, relative to the size of the network. When goods are substitutes, with dense network such that the number of partnerships is large relative to the number of firms operating in the market, the threshold value of the product differentiation parameter can be small.

The use of numerical methods in solving games and dynamic games forms the focus of the final two papers.

In "Global Optimization Approach to Nonzero Sum Six-Person Game," R. Enkhbat and his coauthors examine the nonzero sum six-person game, for which they propose a sufficient condition for a Nash equilibrium based on Mills [1]. The numerical Nash equilibria are obtained using the Curvilinear Multistart Algorithm [2] for nonconvex optimization. This was illustrated using a six-person game and Nash equilibria are found in all cases.

Unfortunately, finding the Nash equilibria is not so trivial for mean field games, especially for an infinite-horizon mean field games. Most mean field games are finite-horizon, since finding the mean field game ϵ-Nash equilibrium can be achieved with the help of the boundary conditions. However, there is no boundary condition for the infinite-horizon mean field games. In "An Infinite-Horizon Mean Field Game of Growth and Capital Accumulation: A Markov Chain Approximation Numerical Scheme and Its Challenges," Chee Kian Leong presents a Markov chain approximation scheme [3] as potentially useful for obtaining the numerical ϵ-Nash equilibrium solution to the infinite-horizon mean field system of equations. However, the implementation of this scheme involves some challenges, such as appropriate choices of initial values for the policy function iteration step and the capital distribution functions in the Fokker–Planck–Kolmogorov forward equation. Consequently, finding an appropriate and robust numerical scheme to solve infinite-horizon mean field games remains a challenging research question in the mean field games literature.

Together, this collection of papers represents the state of the art in games and dynamic games and their applications. Besides being a testimony to the vitality of research in this field, these papers will stimulate further research in both theory and applications of games and dynamic games.

Hong Kong, Hong Kong   David Yeung
Hangzhou, China   Shravan Luckraz
Ningbo, China   Chee Kian Leong

References

1. Mills H. (1960) *Equilibrium Points in Finite Games*. J. Soc. Indust. Appl. Mathemat. 8 (2): 397–402.
2. Gornov A., Zarodnyuk T. (2014) *Computing technology for estimation of convexity degree of the multiextremal function*. Machine learning and data analysis 10 (1): 1345–1353. http://jmlda.org
3. Kushner H.J. (2008) Numerical methods for non-zero-sum stochastic differential games: Convergence of the Markov chain approximation method. In: Chow P.L., Yin, G., Mordukhovich, B. (eds) Topics in Stochastic Analysis and Nonparametric Estimation. The IMA Volumes in Mathematics and its Applications, vol 145. New York, NY: Springer, pp 51–84.

Acknowledgments

We would like to thank the associate editors and all the referees who made this volume possible with their competent advice.

We are grateful to Vladimir Mazalov (President of ISDG), Georges Zaccour (past President of ISDG), HongWei Gao (President of ISDG China Chapter), and Leon Petrosyan and David Yeung (Co-founding Advisor of the China Chapter) for their unfailing support.

We would like to express our gratitude to the Ningbo conference organizing committee—Shravan Luckraz (Chairman), Chee Kian Leong, Qu Chen, Xuewen Qian, and Stephanie Zhao—for making the conference a highly successful event.

Further, Shravan Luckraz is grateful to the National Science Foundation of China (NSFC grant number 71650110521) for its financial support.

Contents

Contributors

Mohamad Alghamdi Mathematics, King Saud University, Riyadh, Saudi Arabia

A. Anikin Matrosov Institute for System Dynamics and Control Theory, SB of RAS, Novosibirsk, Russian Federation

S. Batbileg National University of Mongolia, Ulaanbaatar, Mongolia

R. Enkhbat National University of Mongolia, Ulaanbaatar, Mongolia

Olga I. Gorbaneva J.I. Vorovich Institute of Mathematics, Mechanics and Computer Sciences Southern Federal University, Rostov-on-Don, Russian Federation

Vasily V. Gusev National Research University, Higher School of Economics, St. Petersburg, Russian Federation

A. Gornov Matrosov Institute for System Dynamics and Control Theory, SB of RAS, Novosibirsk, Russian Federation

Chee Kian Leong The University of Nottingham, Ningbo, China

Stuart MacDonald School of Economics, University of Nottingham, Ningbo, China

Hiroaki Mukaidani Graduate School of Engineering, Hiroshima University, Higashi-Hiroshima City, Hiroshima, Japan

Gennady A. Ougolnitsky J.I. Vorovich Institute of Mathematics, Mechanics and Computer Sciences Southern Federal University, Rostov-on-Don, Russian Federation

Bernard Pailthorpe School of Physics, University of Sydney, Sydney, NSW, Australia

Leon A. Petrosyan St. Petersburg State University, Saint Petersburg, Russia

N. Tungalag National University of Mongolia, Ulaanbaatar, Mongolia

David W. K. Yeung Saint Petersburg State University, St Petersburg, Russia
Hong Kong Shue Yan University, North Point, Hong Kong
Department of Finance, Asia University, Taiwan

Part I
Cooperative Dynamic Games
(Plenary Lectures)

Dynamically Stable Cooperative Provision of Public Goods Under Non-transferable Utility

David W. K. Yeung

Abstract The provision of public goods constitutes a classic case of market failure which calls for cooperative optimization. However, cooperation cannot be dynamically stable unless the participating agents' cooperative payoffs are guaranteed to be no less than their non-cooperative payoffs throughout the cooperation duration. This paper derives dynamically stable cooperative solutions for public goods provision by asymmetric agents with non-transferable utility/payoffs in a differential game framework. It examines the dynamically stable subcore under a constant payoff weight. In the case of the absence of a dynamically stable subcore under a constant weight, a variable payoff weights scheme leading to a dynamically stable cooperative solution is designed. This is the first time that dynamically stable cooperative provision of public goods in a non-transferable payoff framework is provided, and further research along this line is expected.

Keywords Public goods · Differential games · Dynamic stability · Non-transferable payoffs

1 Introduction

The provision of public goods constitutes a classic case of market failure which calls for cooperative optimization. The non-exclusive and non-rivalrous properties of public goods constitute a major factor for the failure in their efficient provision by individual parties. Yet the positive externalities of some public goods are rich sources for mutual gains. Examples of such public goods include clean environment,

D. W. K. Yeung
Saint Petersburg State University, St Petersburg, Russia

Hong Kong Shue Yan University, North Point, Hong Kong

Department of Finance, Asia University, Taiwan
e-mail: dwkyeung@hksyu.edu

© Springer Nature Switzerland AG 2020
D. Yeung et al. (eds.), *Frontiers in Games and Dynamic Games*, Annals of the International Society of Dynamic Games 16,
https://doi.org/10.1007/978-3-030-39789-0_1

regional security, scientific knowledge, accessible public capital, transportation infrastructures, technical know-how, and public information. Analysis on provision of public goods in a static framework can be found in Chamberlin [2], McGuire [10], and Gradstein and Nitzan [6]. Given that public goods provision often involves an intertemporal duration, a differential game approach has to be adopted. Studies on voluntary provision of public goods in an intertemporal framework were given by Fershtman and Nitzan [5], Wirl [13], and Wang and Ewald [12]. Dockner et al. [4] presented a differential game model with two asymmetric agents in which knowledge is a public good and focused on the non-cooperative equilibria and the collusive solution that maximizes the joint payoffs of all agents.

Cooperation suggests the possibility of socially efficient solutions to the public goods provision problem. However, cooperation cannot be sustainable unless the participating agents' cooperative payoffs are guaranteed to be not less than their non-cooperative payoffs throughout the cooperation duration. It is due to the lack of this kind of guarantee that many cooperation schemes become unstable and may fail any time within the cooperation duration. A dynamically stable cooperative solution guarantees that participants will always be better off throughout the entire cooperation duration under the agreed-upon optimality principle from the beginning to the end. Yeung and Petrosyan [17] analyzed dynamically stable solutions for provision of public goods when payoffs are transferable. Yeung and Petrosyan [18] considered dynamically stable cooperative provision of public goods under accumulation and payoff uncertainties in a dynamic game framework.

In addition, gains from public or social goods are often non-transferrable. Construction of dynamically stable cooperative solution in the case where payoffs are not transferable is much more difficult than in the case where payoffs are transferable because side payments cannot be used as an instrument. Yeung and Petrosyan [5] and Yeung et al. [21] examined dynamically stable solution in stochastic differential games with non-transferable payoffs/utility. Yeung and Petrosyan [19] presented subgame consistent cooperative solution for non-transferable payoff dynamic games using variable payoff weights. This paper derives dynamically stable cooperative solutions for public goods provision by asymmetric agents with non-transferable payoffs in a differential game framework. The dynamically stable subcore under a constant payoff weight is examined. In the case where the dynamically stable subcore under a constant weight does not exist, a variable payoff weights scheme is presented to obtain a dynamically stable cooperative solution. This is the first time that dynamically stable cooperative provision of public goods in a non-transferable utility framework is provided.

The paper is organized as follows: Sect. 1.2 provides the theoretical framework of cooperative provision of public goods when payoffs/utility is not transferable. In particular, the core, the dynamically stable subcore under constant and variable weights are examined. A differential game of cooperative provision of public goods under non-transferrable utility is presented in Sect. 1.3. The Pareto optimal set is derived and the conditions satisfying individual rationality are presented. Details of a dynamically stable cooperative scheme with a set of constant payoff weights are given in Sect. 1.4. Section 1.5 presents a variable weights cooperative scheme to guarantee the satisfaction of individual rationality in the case where a set of constant weights satisfying individual rationality throughout the entire cooperation duration does not exist. Section 1.6 concludes the paper.

2 Theoretical Framework of Public Goods Provision

Consider the case of n asymmetric agents/nations with a public good. Each agent makes its own contribution to the building up of the stock of the public good. The duration of the planning horizon is $[0, T]$. The payoffs of the agents are not transferrable. We use $K(s)$ to denote the stock of the public good and $I_i(s)$ the public capital investment by nation i at time $s \in [0, T]$. The stock accumulation dynamics is governed by the differential equation:

$$\dot{K}(s) = f [K(s), I_1(s), I_2(s), \cdots, I_n(s), s] \ K(0) = K_0. \tag{2.1}$$

The instantaneous payoff to agent i at time instant s is

$$u^i \left\{ R_i^{(1)} \left[q_i^1(s), (K(s)), s \right], R_i^{(2)} \left[q_i^2(s), s \right], c_i [I_i(s), s] \right\}$$

which includes the benefits from the public goods in productive activities and from the public good itself—$R_i^{(1)} \left[q_i^1(s), (K(s)), s \right]$, the benefit to the agent in productive activities not involving the public good—$R_i^{(2)} \left[q_i^2(s), s \right]$, and investment costs in the public good—$c_i[I_i(s), s]$. In particular, $q_i^1(s)$ are the productive activities which depend on the public good, and $q_i^2(s)$ are the productive activities independent of the public good.

The objective of agent $i \in N$ is to maximize its payoff over the planning horizon T.

$$\int_0^T u^i \left\{ R_i^{(1)} \left[q_i^1(s), (K(s)), s \right], R_i^{(2)} \left[q_i^2(s), s \right], c_i [I_i(s), s] \right\} e^{-rs} ds$$
$$+ m_i^1 [K(T)] e^{-rT}, \tag{2.2}$$

subject to the public capital stock accumulation dynamics (2.1), where r is the discount rate and $m_i^1 [K(T)]$ is the terminal payoff at time T.

We use $V^i(t, K)$ to denote the non-cooperative game equilibrium payoff of agent $i \in N$ and $\{q_i^{1*}(s) = \phi_i^1(s, K), q_i^{2*}(s) = \phi_i^2(s, K), I_i^*(s) = \phi_i^3(s, K),$ for $i \in N$ and $s \in [0, T]\}$ to denote the game equilibrium strategies.

2.1 Pareto Optimality and Individual Rationality under Cooperation

It is a well-known problem that non-cooperative provision of goods with externalities, in general, would lead to inefficiency. Cooperation suggests the possibility of socially efficient solutions. We first consider the case where the agents agree to set up a cooperative scheme in public goods provision with a constant payoff weight. Under cooperation, the agents negotiate to establish an agreement (optimality principle) on the level of investment in public goods production provided by each agent. In a non-transferable payoffs framework, the agents' imputations/payoffs are directly dictated by the agreed-upon cooperative investment strategies. A necessary condition for dynamic stability is that the agreed-upon optimality principle must satisfy individual rationality throughout the cooperation duration.

To obtain the core of the non-transferrable utility (NTU) cooperative dynamic public goods provision game (2.1)–(2.2), we first derive the Pareto optimal trajectories by maximizing the weighted sum of payoffs of the agents under different payoff weights (see Leitmann [8], Dockner and Jørgensen [3], Hamalainen et al. [7], Yeung and Petrosyan [15, 20] and Yeung et al. [21]). Consider the case in which the agents adopt a payoff weight vector $\alpha = (\alpha^1, \alpha^2, \cdots, \alpha^n)$ in all stages, where $\sum_{j=1}^{n} \alpha^j = 1$.

Conditional upon the weight α, the agents' optimal cooperative strategies can be generated by solving the dynamic programming problem of maximizing the weighted sum of payoffs:

$$
\sum_{j=1}^{n} \alpha_j \left(\int_0^T u^j \left\{ R_j^{(1)} \left[q_j^1(s), (K(s)), s \right], R_j^{(2)} \left[q_j^2(s), s \right], c_j \left[I_j(s), s \right] \right\} e^{-rs} ds \right.
$$

$$
\left. + m_i^1 \left[K(T) \right] e^{-rT} \right) \tag{2.3}
$$

subject to (2.1).

We use $W^{(\alpha)}(t, K)$ to denote the maximized weighted sum of payoffs under cooperation, and $\{ q_i^{1*}(s) = \psi_i^{(\alpha)1}(s, K), q_i^{2*}(s) = \psi_i^{(\alpha)2}(s, K), I_i^*(s) = \psi_i^{(\alpha)3}(s, K)$, for $i \in N$ and $s \in [0, T]\}$ to denote the cooperative strategies under payoff weight α. Substituting the cooperative investment strategies $\left\{ \psi_i^{(\alpha)3}(s, K) \right\}$ into the state dynamics (2.1), one can obtain the dynamics of the cooperative trajectory as:

$$
\dot{K}(s) = f \left[K(s), \psi_1^{(\alpha)3}(s, K), \psi_2^{(\alpha)3}(s, K), \cdots, \psi_n^{(\alpha)3}(s, K), s \right], K(0) = K_0. \tag{2.4}
$$

We use $\left\{ K^{(\alpha)}(s) \right\}_{s=0}^{T}$ to denote the solution path of (2.4). We also use $K_s^{(\alpha)} = K^{(\alpha)}(s)$ interchangeably if there is no ambiguity.

2.2   The Core and Dynamically Stable Subcore

The core $\overline{C}\,(0, K_0)$ of the NTU cooperative dynamic game (2.1)–(2.2) is the set of imputations:

$\{W^{(\alpha)i}(0, K_0), \text{ for } i \in N\}$ such that.

$W^{(\alpha)i}(0, K_0)$ satisfies Pareto optimality and the individual rationality conditions that

$$W^{(\alpha)i}\,(0, K_0) \geq V^i\,(0, K_0),\, \text{for } i \in N. \tag{2.5}$$

Note that non-cooperative equilibria are sub-optimal in general. Therefore, there always exists a non-empty core $\overline{C}\,(0, K_0)$, even if the non-cooperative equilibrium outcome is the only element. The core $\overline{C}\,(0, K_0)$ does not guarantee that individual rationality will be satisfied throughout the entire cooperation duration, that is, the condition $W^{(\alpha)i}\left(t, K_t^{(\alpha)}\right) \geq V^i\left(t, K_t^{(\alpha)}\right)$, for $i \in N$, $t \in [0, T]$. We use $\overline{\Lambda}$ to denote the set of payoff weights in the core. Frequently, the lack of sustainability of the cooperation scheme leads to break-up of the scheme as the game evolves. To overcome the problem, one has to consider a subcore which is dynamically stable. Let α be an element in $\overline{\Lambda}$ such that.

$$W^{(\alpha)i}\left(t, K_t^{(\alpha)}\right) \geq V^i\left(t, K_t^{(\alpha)}\right),\, \text{for } i \in N \text{ and } t \in [0, T]. \tag{2.6}$$

We use Λ to denote the set of α that satisfies (2.6).

The dynamically stable subcore $C\left(t, K_t^{(\alpha)}\right)$ of the NTU cooperative dynamic game (2.1)–(2.2) is the set of imputations:

$\left\{W^{(\alpha)i}\left(t, K_t^{(\alpha)}\right), \text{ for } i \in N \text{and } t \in [0, T]\right\}$ such that (2.6) is satisfied.

If Λ is not an empty set, a weight $\hat{\alpha} \in \Lambda$ agreed upon by all agents would yield a cooperative solution which satisfies both individual rationality and Pareto optimality throughout the cooperation duration. However, a dynamically stable subcore $C\left(t, K_t^{(\alpha)}\right)$ may be empty.

2.3   Dynamically Stable Subcore with Variable Weights

To resolve the problem of an empty dynamically stable constant weight core, time varying payoff weights can be adopted. Sorger [11], Marin-Solano [9], and Yeung and Petrosyan [19] considered the derivation of solutions for NTU cooperative dynamic games with variable payoff weights.

To derive a dynamically stable scheme for cooperative provision of public goods under non-constant payoff weights, the agents have to agree/bargain on a specific pattern of partitions of the game horizon. Let the agreed-upon partitions of the game horizon be denoted by the τ partitions: $[0, t_1), [t_1, t_2), [t_2, t_3), \cdots, [t_{\tau - 1}, T]$. In each of these partitions, there exists a non-empty set of constant payoff weight Λ^k, for

$k \in \{1, 2, \cdots, \tau\}$, which satisfies individual rationality. Then, the agents seek an agreed-upon weight $\hat{\alpha}^k \in \Lambda^k$ in the time interval $[t_{k-1}, t_k)$, for $k \in \{1, 2, \cdots, \tau\}$.

The pros of the variable payoff weights scheme include the flexibility in accommodating the preferences of the agents according to the initial cooperative agreement and the resolution of the problem of an empty dynamically stable constant weight core. On the other hand, the cons thereof is that full Pareto efficiency is sacrificed for obtaining a dynamically stable solution.

3   An NTU Differential Game of Public Goods Provision

In this section, we present a cooperative game of public goods provision with non-transferable payoffs. A group of n agents makes continuous contributions of some inputs or investments to build up a productive stock of a public good. We use $K(s)$ to denote the level of the productive stock and $I_i(s)$ the contribution or investment by agent i at time s, the stock accumulation dynamics is

$$\dot{K}(s) = f[K(s), I_1(s), I_2(s), \cdots, I_n(s), s] = \sum_{j=1}^{n} I_j(s) - \delta K(s), \ K(0) = K_0,$$

$$(3.1)$$

where δ is the rate of decay of the productive stock.

The benefit to the agent in productive activities including the public good is

$$R_i^{(1)}\left[q_i^1(s), (K(s)), s\right] = \left[a_i e^{\varpi_i s} q_i^1(s)(K(s))^{1/2} - \left(q_i^1(s)\right)^2 + e^{\varsigma_i s} g_i K(s)\right].$$

The benefit to the agent in productive activities not including the public good is

$$R_i^{(2)}\left[q_i^2(s), s\right] = \left[b_i e^{\omega_i s} q_i^2(s) - \left(q_i^2(s)\right)^2\right].$$

The investment costs in the public good is

$$C_i[I_i(s), s] = c_i e^{\kappa_i s} [I_i(s)]^2.$$

The instantaneous payoff to agent i at time instant s is

$$\left[a_i e^{\varpi_i s} q_i^1(s)(K(s))^{1/2} - \left(q_i^1(s)\right)^2 + e^{\varsigma_i s} g_i K(s)\right] + \xi_i \left[b_i e^{\omega_i s} q_i^2(s) - \left(q_i^2(s)\right)^2\right]$$

$$-c_i e^{\kappa_i s} [I_i(s)]^2, i \in N,$$

where ξ_i is a weight reflecting the relative importance of benefits of the two productive activities to agent i.

The term $e^{\varpi_i s}$ exhibits the growth/decline of the sector involving the public good. The term $e^{\varsigma_i s}$ exhibits the growth/decline of the direct impact of the public good. The term $e^{\omega_i s}$ exhibits the growth/decline of the sector not involving the public good. The term $e^{\kappa_i s}$ reflects the change in the cost of investment in the public good over time. The payoffs of the agents are not transferrable.

The objective of agent i is to maximize its payoff over the planning horizon

$$\int_0^T \left\{ \left[a_i e^{\varpi_i s} q_i^1(s)(K(s))^{1/2} - \left(q_i^1(s) \right)^2 + e^{\varsigma_i s} g_i K(s) \right] + \xi_i \left[b_i e^{\omega_i s} q_i^2(s) - \left(q_i^2(s) \right)^2 \right] \right.$$

$$\left. - c_i e^{\kappa_i s} [I_i(s)]^2 \right\} e^{-rs} ds + \left[m_i^1 K(T) + m_i^2 \right] e^{-rT}, \text{ for } i \in N, \qquad (3.2)$$

subject to the public capital stock accumulation dynamics (3.1),

where r is the rate of time preference, and $\left[m_i^1 K(T) + m_i^2 \right]$ is agent i's terminal payoff conditional on the stock of public capital at time T.

Acting on self-interests, the agents are involved in a differential game. In such a framework, a feedback Nash equilibrium has to be sought. Invoking the standard techniques for solving differential games, a non-cooperative feedback solution to the game (3.1)–(3.2) can be characterized as follows: A set of feedback strategies $\left\{ q_i^{1*}(s) = \phi_i^1(s, K), q_i^{2*}(s) = \phi_i^2(s, K), I_i^*(s) = \phi_i^3(s, K), \text{ for } i \in N \text{ and } s \in [0, T] \right\}$ constitutes a feedback Nash equilibrium of the game (3.1)–(3.2) if there exists differentiable functions $V^i(t, K) : [0, T] \times R \to R$ for $i \in N$ satisfying the following set of Hamilton–Jacobi–Bellman equations (see Basar and Olsder [1]; Yeung and Petrosyan [16, 20])

$$-V_t^i(t, K) = \max_{q_i^1, q_i^2, I_i} \left\{ \left\{ \left[a_i e^{\varpi_i t} q_i^1 K^{1/2} - \left(q_i^1 \right)^2 + e^{\varsigma_i t} g_i K \right] + \xi_i \left[b_i e^{\omega_i t} q_i^2 - \left(q_i^2 \right)^2 \right] \right. \right.$$

$$\left. \left. - c_i e^{\kappa_i t} (I_i)^2 \right\} e^{-rs} + V_K^i(t, K) \left[\sum_{\substack{j=1 \\ j \neq i}}^n \phi_j^3(t, K) + I_i - \delta K \right] \right\}, \qquad (3.3)$$

$$V^i(T, K) = \left[m_i^1 K + m_i^2 \right] e^{-rT}, \text{ for } i \in N. \qquad (3.4)$$

Performing the indicated maximization in (3.3) yields the game equilibrium strategies:

$$q_i^{1*}(t) = \phi_i^1(t, K) = \left(a_i e^{\varpi_i t}\right) K^{1/2}/2,$$

$$q_i^{2*}(t) = \phi_i^2(t, K) = \left(b_i e^{\omega_i t}\right)/2,$$

$$I_i^*(t) = \phi_i^3(t, K) = V_K^i(t, K) e^{rt}/2c_i e^{\kappa_i t}, \text{ for } i \in N \tag{3.5}$$

The payoff of the agents under non-cooperation can be obtained as:

Proposition 3.1

$$V^i(t, K) = [A_i(t)K + C_i(t)] e^{-rt}, \text{ for } i \in N \text{ and } t \in [0, T]; \tag{3.6}$$

where the value of $A_i(t)$ is generated by the first-order differential equation:

$$\dot{A}_i(t) = (r + \delta) A_i(t) - \frac{\left(a_i e^{\varpi_i t}\right)^2}{4} - e^{\varsigma_i t} g_i, \ A_i(T) = m_i^1;$$

and the value of $C_i(t)$ is generated by the following first-order differential equations:

$$\dot{C}_i(t) = rC_i(t) + \frac{[A_i(t)]^2}{4c_i e^{2\kappa_i t}} - \left[\sum_{j=1}^n \frac{A_i(t)A_j(t)}{2c_j e^{\kappa_j t}} \right] - \xi_i \frac{\left(b_i e^{\omega_i t}\right)^2}{4},$$

$$C_i(T) = m_i^2, \text{ for } i \in N.$$

Proof See Appendix A.

4 Cooperative Scheme: Constant Payoff Weight Case

Consider the case in which the agents agree to set up a cooperative scheme in public goods provision with a constant payoff weight.

4.1 Optimal Cooperative Trajectories

To derive a set of optimal cooperative strategies the agents have to agree on a payoff weight $\alpha = (\alpha_1, \alpha_2, \cdots, \alpha_n)$ for $\sum_{j=1}^n \alpha_j = 1$. Conditional upon an agreed-upon weight α, the agents' optimal cooperative strategies can be generated by solving the following optimal control problem (See Leitmann [8], Yeung and Petrosyan [15, 20] and Yeung et al. [21]):

$$\sum_{j=1}^{n} \alpha_j \left(\int_0^T \left\{ \left[a_j e^{\varpi_j s} q_j^1(s)(K(s))^{1/2} - \left(q_j^1(s) \right)^2 + e^{\varsigma_j s} g_j K(s) \right] \right. \right.$$

$$+ \xi_j \left[b_j e^{\omega_i s} q_j^2(s) - \left(q_j^2(s) \right)^2 \right]$$

$$\left. \left. - c_j e^{\kappa_j s} [I_j(s)]^2 \right\} e^{-rs} ds + \left[m_j^1 K(T) + m_j^2 \right] e^{-rT} \right), \qquad (4.1)$$

subject to dynamics (3.1).

Invoking the standard dynamic programming technique an optimal solution to the control problem (3.1) and (4.1) can be characterized as follows. A set of controls $\left\{ q_i^{1*}(s) = \psi_i^{1(\alpha)}(s, K), q_i^{2*}(s) = \psi_i^{2(\alpha)}(s, K), I_i^*(s) = \psi_i^{3(\alpha)}(s, K), \text{ for } i \in N \text{ and } s \in [0, T] \right\}$ constitutes an optimal solution to the problem (3.1). (4.1) if there exist differentiable functions $W_i^{(\alpha)}(t, K) : [0, T] \times R \to R$ for $i \in N$ satisfying the following set of partial differential equations (see Basar and Olsder [1]; Yeung and Petrosyan [16, 20]):

$$-W_t^{(\alpha)}(t, K) = \max_{\substack{q_i^1, q_i^2, I_i, \\ i \in N}} \left\{ \sum_{j=1}^{n} \alpha_j \left\{ \left[a_j e^{\varpi_j t} q_j^1 K^{1/2} - \left(q_j^1 \right)^2 + e^{\varsigma_j t} g_j K \right] \right. \right.$$

$$+ \xi_j \left[b_j e^{\omega_i t} q_j^2 - \left(q_j^2 \right)^2 \right]$$

$$\left. \left. - c_j e^{\kappa_j s} \left(I_j \right)^2 \right\} e^{-rs} + W_K^{(\alpha)}(t, K) \left[\sum_{j=1}^{n} I_j - \delta K \right] \right\}, \qquad (4.2)$$

$$W^{(\alpha)}(T, K) = \sum_{j=1}^{n} \alpha_j \left[m_j^1 K + m_j^2 \right] e^{-rT}. \qquad (4.3)$$

Performing the indicated maximization in (4.2) yields

$$q_i^{1*}(t) = \psi_i^{1(\alpha)}(t, K) = \left(a_i e^{\varpi_i t} \right) K^{1/2}/2,$$

$$q_i^{2*}(t) = \psi_i^{2(\alpha)}(t, K) = \left(b_i e^{\omega_i t} \right)/2,$$

$$I_i^*(t) = \psi_i^{3(\alpha)}(t, K) = W_K^{(\alpha)}(t, K) e^{rt}/2c_i e^{\kappa_i t} \alpha_i, \text{ for } i \in N. \qquad (4.4)$$

Condition (4.4) reflects that an optimum investment will be made up to the point where the magnitude of the product of the marginal disutility of investment cost and the marginal cost of investment of agent i equals to the implicit marginal benefit of the productive stock to all the participating agents divided by agent i's assigned weight α_i. Therefore, the lower the assigned weight the higher level of public goods investment, other things being equal.

The value function reflecting the maximized weight sum of the agents' payoff $W^{(\alpha)}(t, K)$ in the interval $[t, T]$ can be obtained as:

Proposition 4.1

$$W^{(\alpha)}(t, K) = \left[A^{(\alpha)}(t)K + C^{(\alpha)}(t) \right] e^{-rt}, \text{ for } t \in [t_1, T]; \qquad (4.5)$$

where the value of $A^{(\alpha)}(t)$ is generated by the first-order differential equation:

$$\dot{A}^{(\alpha)}(t) = (r+\delta) A^{(\alpha)}(t) - \sum_{j=1}^{n} \alpha_j \left[\frac{\left(a_j e^{\varpi_j t}\right)^2}{4} + e^{\varsigma_j t} g_j \right], \quad A^{(\alpha)}(T) = \sum_{j=1}^{n} \alpha_j m_j^1;$$

and the value of $C^{(\alpha)}(t)$ is generated by the first-order differential equation:

$$\dot{C}^{(\alpha)}(t) = rC^{(\alpha)}(t) + \sum_{j=1}^{n} \frac{\left[A^{(\alpha)}(t)\right]^2}{4\alpha_j c_j e^{2\kappa_j t}} - \sum_{j=1}^{n} \frac{\left[A^{(\alpha)}(t)\right]^2}{2\alpha_j c_j e^{\kappa_j t}} - \sum_{j=1}^{n} \frac{\alpha_j \xi_j \left(b_j e^{\omega_i s}\right)^2}{4},$$

$$C^{(\alpha)}(T) = \sum_{j=1}^{n} \alpha_j m_j^2.$$

Proof See Appendix B.

Substituting the optimal public good strategies $\left\{ \psi_i^{3(\alpha)}(s, K), \text{ for } i \in N \text{ and } s \in [0, T] \right\}$ into (3.1) yields the optimal path of the productive stock dynamics:

$$\dot{K}(s) = \left[\sum_{j=1}^{n} \frac{A^{(\alpha)}(s)}{2c_j e^{\kappa_j s} \alpha_j} - \delta K(s) \right], \quad K(0) = K_0. \qquad (4.6)$$

We use $\left\{ K^{(\alpha)}(s) \right\}_{s=0}^{T}$ to denote the solution generated by (4.6). The terms $K^{(\alpha)}(s)$ and $K_s^{(\alpha)}$ are used interchangeably.

4.2 Individual Payoff Under Cooperation

In the cooperation duration $s \in [0, T]$, agents will use the cooperative strategies.

$$q_i^{1*}(s) = \psi_i^{1(\alpha)}(s, K) = \left(a_i e^{\varpi_i s}\right) K^{1/2}/2,$$

$$q_i^{2*}(s) = \psi_i^{2(\alpha)}(s, K) = \left(b_i e^{\varpi_i s}\right)/2,$$

$$I_i^*(s) = \psi_i^{3(\alpha)}(s, K) = A_s^{(\alpha)}(s)/2c_i e^{\kappa_i s}\alpha_i, \text{ for } i \in N. \tag{4.7}$$

The payoff of the agent i over the duration $[t, T]$ under cooperation given that the level of public capital stock is K can be obtained as:

$$W^{(\alpha)i}(t, K) = \int_t^T \left[\frac{(a_i e^{\varpi_i s})^2 K(s)}{4} + e^{\varsigma_i s} g_i K(s) + \xi_i \frac{(b_i e^{\varpi_i s})^2}{4} - \frac{[A^{(\alpha)}(s)]^2}{4c_i e^{2\kappa_i s}(\alpha_i)^2} \right] e^{-rs} ds$$

$$+ \left[m_i^1 K(T) + m_i^2 \right] e^{-rT}, \text{ for } i \in N \text{ and } t \in [0, T] \bigg\}. \tag{4.8}$$

Following Yeung [14], we can use a set of partial differential equations similar to the Hamilton–Jacobi–Bellman equations to characterize the solution of the function $W^{(\alpha)i}(t, K)$ which allows its derivation in a more direct way.

Proposition 4.2 *The value function* $W^{(\alpha)i}(t, K)$ *can be characterized as:*

$$-W_t^{(\alpha)i}(t, K) = \left[\frac{\left(a_i e^{\varpi_i t}\right)^2 K}{4} + e^{\varsigma_i t} g_i K + \xi_i \frac{(b_i e^{\omega_i t})^2}{4} - \frac{W_K^{(\alpha)i}(t, K)}{4c_i e^{2\kappa_i t}(\alpha_i)^2} \right] e^{-rt}$$

$$+ W_K^{(\alpha)i}(t, K) \left[\sum_{j=1}^n \frac{A^{(\alpha)}(t)}{2c_j e^{\kappa_j t}\alpha_j} - \delta K \right], \tag{4.9}$$

$$W^{(\alpha)i}(T, K) = \left[m_i^1 K + m_i^2 \right] e^{-rT}, \text{ for } i \in N \text{ and } t \in [0, T]. \tag{4.10}$$

Proof See Yeung [14].

Therefore, if there exist differentiable functions $W^{(\alpha)i}(t, K) : [0, T] \times R \to R$ for $i \in N$ satisfying (4.9)–(4.10), the function $W^{(\alpha)i}(t, K)$ yields the payoff of agent i under the cooperative scheme.

Proposition 4.3 *The value function* $W^{(\alpha)i}(t, K)$ *can be obtained as:*

$$W^{(\alpha)i}(t, K) = \left[A_i^{(\alpha)}(t)K + C_i^{(\alpha)}(t) \right] e^{-rt}, \text{ for } i \in N \text{ and } t \in [0, T]; \tag{4.11}$$

where the value of $A_i^{(\alpha)}(t)$ is generated by the following first-order differential equation:

$$\dot{A}_i^{(\alpha)}(t) = (r + \delta) A_i^{(\alpha)}(t) - \frac{\left(a_i e^{\varpi_i t}\right)^2}{4} - e^{\varsigma_i t} g_i, \ A_i(T) = m_i^1;$$

and the value of $C_i^{(\alpha)}(t)$ is generated by the following first-order differential equations:

$$\dot{C}_i^{(\alpha)}(t) = r C_i^{(\alpha)}(t) + \frac{A_i^{(\alpha)}(t, K)}{4 c_i e^{2\kappa_i t}(\alpha_i)^2} - A_i^{(\alpha)} \sum_{j=1}^{n} \frac{A^{(\alpha)}(t)}{2\alpha_j c_j e^{\kappa_j t}} - \xi_i \frac{\left(b_i e^{\varpi_i t}\right)^2}{4},$$

$$C_i^{(\alpha)}(T) = m_i^2.$$

Proof See Appendix C.

4.3 Dynamically Stable Core

An essential element for a cooperative scheme to be acceptable to all agents is the satisfaction of individual rationality, that is, the payoff of each agent under cooperation is no less than that under non-cooperation. In particular, at the outset

$$W^{(\alpha)i}(0, K_0) \geq V^i(0, K_0), \text{ for } i \in N.$$

For stability in a dynamic framework, individual rationality has to be satisfied throughout the cooperation duration along the cooperative state trajectory. Hence dynamic stability implies

$$W^{(\alpha)i}\left(t, K_t^{(\alpha)}\right) = \left[A_i^{(\alpha)}(t) K_t^{(\alpha)} + C_i^{(\alpha)}(t)\right] e^{-rt} \geq V^i\left(t, K_t^{(\alpha)}\right)$$

$$= \left[A_i(t) K_t^{(\alpha)} + C_i(t)\right] e^{-rt},$$

$$\text{for } i \in N \text{ and } t \in [0, T]. \tag{4.12}$$

If condition (4.12) is not satisfied for an agent at any time instant within the cooperation duration, he would opt out of the cooperative scheme and act independently.

To derive the set of payoff weights which are candidates for weights leading to a dynamically stable solution, we have to search the set of vectors α such that:

$$W^{(\alpha)i}(t, K) = \left[A_i^{(\alpha)}(t)K + C_i^{(\alpha)}(t)\right]e^{-rt} \geq V^i(t, K) = [A_i(t)K + C_i(t)]e^{-rt},$$

for $i \in N$ and $t \in [0, T]$.

Invoking Proposition 3.1 and Proposition 4.2, the difference between $W^{(\alpha)i}(t, K)$ and $V^i(t, K)$ for agent i can be expressed as:

$$W^{(\alpha)i}(t, K) - V^i(t, K) = \left[C_i^{(\alpha)}(t) - C_i(t)\right]e^{-rt}, \text{ for } i \in N \text{ and } t \in [0, T].$$

$$(4.13)$$

We use Λ to define the set of weight vectors $\alpha = (\alpha_1, \alpha_2, \cdots, \alpha_n)$ such that.

$$W^{(\alpha)i}\left(t, K_t^{(\alpha)}\right) \geq V^i\left(t, K_t^{(\alpha)}\right), \text{ for } i \in N \text{ and } t \in [0, T]. \qquad (4.14)$$

Note that the set Λ can be an empty set. If Λ is not an empty set, then there exists a dynamically stable core of the cooperative game (3.1)–(3.2). A vector $\hat{\alpha} = (\hat{\alpha}_1, \hat{\alpha}_2, \cdots, \hat{\alpha}_n) \in \Lambda$ agreed upon by all agents would yield a dynamically stable solution.

Given the closed-form solutions of the functions $W^{(\alpha)i}\left(t, K_t^{(\alpha)}\right)$ and $V^i\left(t, K_t^{(\alpha)}\right)$, we search for the existence of a dynamically stable core. It was found that time invariant growth parameters—$a_j e^{\varpi_j s} = a_j$, $b_j e^{\varpi_j s} = b_j$, and $c_j e^{\kappa_j s} = c_j$, for all agents $j \in N$—generally lead to the case of having a dynamically stable core. The non-existence of a dynamically stable core arises in some cases where the growth rates ϖ_j and κ_j for $j \in N$ are significantly different from ϖ_ℓ and κ_ℓ for $\ell \in N$ and $j \neq \ell$.

5 Variable Payoff Weights Scheme

It is possible that there does not exist any constant weight such that the individual rationality condition in (4.12) is satisfied throughout the game duration. An alternative scheme is to search for a set of variable weights over the cooperation duration under which individual rationality is to be maintained at all instants of time. Adopting variable weights entails a trade-off of Pareto optimality for individual rationality.

To derive a dynamically stable scheme for cooperative provision of public goods under non-constant payoff weights the agents adopt the τ agreed-upon partitions: $[0, t_1), [t_1, t_2), [t_2, t_3), \cdots, [t_{\tau-1}, T]$. In each of these partitions, there exists a non-empty set of constant payoffs weight Λ^k, for $k \in \{1, 2, \cdots, \tau\}$, which satisfies individual rationality.

In the cooperative subgame in the last interval $[t_{\tau-1}, T]$, to achieve Pareto optimality the agents jointly maximize the weighted sum of payoffs.

$$\sum_{j=1}^{n} \alpha_j^{\tau} \left(\int_{t_{\tau-1}}^{T} \{ \left[a_j e^{\varpi_j s} q_j^1(s)(K(s))^{1/2} - \left(q_j^1(s) \right)^2 + e^{\varsigma_j t} g_j K(s) \right] \right.$$

$$+\xi_j \left[b_j e^{\omega_i s} q_j^2(s) - \left(q_j^2(s) \right)^2 \right]$$

$$\left. -c_j e^{\kappa_j s} \left[I_j(s) \right]^2 \} e^{-rs} ds + \left[m_j^1 K(T) + m_j^2 \right] e^{-rT} \right) \tag{5.1}$$

subject to dynamics (3.1).

Note that there exists a set of weights $\alpha^{\tau} = \left(\alpha_1^{\tau}, \alpha_2^{\tau}, \cdots, \alpha_n^{\tau} \right) \in \Lambda^{\tau}$ satisfying $W^{(\alpha^{\tau})i}(t, K) \geq V^i(t, K)$, for $i \in N$ and $t \in [t_{\tau-1}, T]$. Let the agreed-upon weight for the interval $[t_{\tau-1}, T]$ be $\hat{\alpha}^{\tau} \in \Lambda^{\tau}$. Invoking Proposition 4.3, one can obtain individual agents' payoffs under cooperation as:

$$W^{(\hat{\alpha}^{\tau})i}(t, K) = \left[A_i^{(\hat{\alpha}^{\tau})}(t) K + C_i^{(\hat{\alpha}^{\tau})}(t) \right] e^{-rt}, \text{ for } i \in N \text{ and } t \in [t_{\tau-1}, T].$$

Now consider the game in the time interval $[t_{\tau-2}, t_{\tau-1})$. The terminal payoff of agent i at time $t_{\tau-1}$ becomes $W^{(\hat{\alpha}^{\tau})i}(t_{\tau-1}, K) = \left[A_i^{(\hat{\alpha}^{\tau})}(t_{\tau-1}) K(t_{\tau-1}) + C_i^{(\hat{\alpha}^{\tau})}(t_{\tau-1}) \right] e^{-rt_{\tau-1}}$ for $i \in N$. Under cooperation the agents jointly maximize the weighted sum of payoffs.

$$\sum_{j=1}^{n} \alpha_j^{\tau-1} \left(\int_{t_{\tau-2}}^{t_{\tau-1}} \{ \left[a_j e^{\varpi_j s} q_j^1(s)(K(s))^{1/2} - \left(q_j^1(s) \right)^2 + e^{\varsigma_j t} g_j K(s) \right] \right.$$

$$+\xi_j \left[b_j e^{\omega_i s} q_j^2(s) - \left(q_j^2(s) \right)^2 \right] - c_j e^{\kappa_j s} \left[I_j(s) \right]^2 \} e^{-rs} ds$$

$$\left. + \left[A^{(\hat{\alpha}^{\tau})}(t_{\tau-1}) K(t_{\tau-1}) + C^{(\hat{\alpha}^{\tau})}(t_{\tau-1}) \right] e^{-rt_{\tau-1}} \right) \tag{5.2}$$

subject to dynamics (3.1).

Again in the time interval $[t_{\tau-2}, t_{\tau-1})$ there exists a set of weights $\alpha^{\tau-1} \in \Lambda^{\tau-1}$ such that $W^{(\alpha^{\tau-1};\hat{\alpha}^{\tau})i}(t, K_t) \geq V^i(t, K_t)$, for $i \in N$ and $t \in [t_{\tau-2}, t_{\tau-1})$. Let the agreed-upon weight in the time interval $[t_{\tau-2}, t_{\tau-1})$ be $\hat{\alpha}^{\tau-1} \in \Lambda^{\tau-1}$. Invoking Proposition 4.3, one can obtain individual agents' payoffs under cooperation as:

$$W^{(\hat{\alpha}^{\tau-1};\hat{\alpha}^{\tau})i}(t, K) = \left[A_i^{(\hat{\alpha}^{\tau-1};\hat{\alpha}^{\tau})}(t)K + C_i^{(\hat{\alpha}^{\tau-1};\hat{\alpha}^{\tau})}(t) \right] e^{-rt},$$

for $i \in N$ and $t \in [t_{\tau-2}, t_{\tau-1}]$.

Following the above analysis, the cooperative subgame in the time interval $[t_{k-1}, t_k)$ for $k \in \{1, 2, \cdots, \tau - 3\}$ can be formulated as maximizing:

$$\sum_{j=1}^{n} \alpha_j^k \left(\int_{t_{k-1}}^{t_k} \{ \left[a_j e^{\varpi_j s} q_j^1(s)(K(s))^{1/2} - \left(q_j^1(s) \right)^2 \right] + \xi_j \left[b_j e^{\omega_i s} q_j^2(s) - \left(q_j^2(s) \right)^2 \right] \right.$$

$$- c_j e^{\kappa_j s} \left[I_j(s) \right]^2 \} e^{-rs} ds$$

$$+ \left[A^{(\hat{\alpha}^{k+1};\hat{\alpha}^{k+2},\hat{\alpha}^{k+3},\cdots,\hat{\alpha}^{\tau})}(t_k) K(t_k) + C^{(\hat{\alpha}^{k+1};\hat{\alpha}^{k+2},\hat{\alpha}^{k+3},\cdots,\hat{\alpha}^{\tau})}(t_k) \right] e^{-rt_k} \right),$$

(5.3)

subject to dynamics (3.1).

Again in the time interval $[t_{k-1}, t_k)$ there exists a set of weights $\alpha^k \in \Lambda$ such that $W^{(\alpha^k;\hat{\alpha}^{k+1},\hat{\alpha}^{k+2},\cdots,\alpha^{\tau})i}(t, K) \geq V^i(t, K)$, for $i \in N$ and $t \in [t_{k-1}, t_k)$. Let the agreed-upon weight in the time interval $[t_{k-1}, t_k)$ be $\hat{\alpha}^k \in \Lambda^k$. Invoking Proposition 4.3, one can obtain individual agents' payoffs under cooperation as:

$$W^{(\hat{\alpha}^k;\hat{\alpha}^{k+1},\hat{\alpha}^{k+2},\cdots,\alpha^{\tau})i}(t, K)$$

$$= \left[A_i^{(\hat{\alpha}^k;\hat{\alpha}^{k+1},\hat{\alpha}^{k+2},\cdots,\alpha^{\tau})}(t)K + C_i^{(\hat{\alpha}^k;\hat{\alpha}^{k+1},\hat{\alpha}^{k+2},\cdots,\alpha^{\tau})}(t) \right] e^{-rt},$$

for $i \in N, t \in [t_{k-1}, t_k)$, and $k \in \{1, 2, \cdots, \tau - 3\}$.

Thus, the agents seek agreed-upon weights $\hat{\alpha}^k \in \Lambda^k$ in the time interval $[t_{k-1}, t_k)$, for $k \in \{1, 2, \cdots, \tau\}$. The cooperative scheme for public goods provision is guided by the payoff weights $(\hat{\alpha}^1, \hat{\alpha}^2, \cdots, \hat{\alpha}^\tau)$ with optimal cooperative strategies:

$$\left\{ \psi_i^{1(\hat{\alpha}^k;\hat{\alpha}^{k+1},\hat{\alpha}^{k+2},\cdots,\alpha^\tau)}(s, K), \psi_i^{2(\hat{\alpha}^k;\hat{\alpha}^{k+1},\hat{\alpha}^{k+2},\cdots,\alpha^\tau)}(s, K), \right.$$

$$\left. \psi_i^{3(\hat{\alpha}^k;\hat{\alpha}^{k+1},\hat{\alpha}^{k+2},\cdots,\alpha^\tau)}(s, K) \right\},$$

for $i \in N, s \in [t_{k-1}, t_k)$, and $k \in \{1, 2, \cdots, \tau\}$.

6   Concluding Remarks

Although cooperative provision of public goods is the key to a socially efficient
solution, one may find it hard to be convinced that dynamic cooperation can offer
a long-term solution unless every participating agent will realize a payoff under
cooperation which is no less than that under non-cooperation from the beginning
to the end. In the case where payoffs are non-transferable the formulation of
dynamically stable cooperative schemes becomes more difficult as side payments
are not possible. This paper resolves the classical problem of market failure in the
provision of public goods with a cooperative scheme in a non-transferable payoffs
environment. Various further research and applications are expected.

Acknowledgements Financial support from the SRS Consortium for Advanced Study in Dynamic
Cooperative Games is gratefully acknowledged.

A.1   Proof of Proposition 3.1

Using Proposition 3.1 and the game equilibrium strategies (3.5) the set of Hamilton–
Jacobi–Bellman equations in (3.4) can be expressed as:

$$r\left[A_i^{(\alpha)}(t)K + C_i^{(\alpha)}(t)\right] - \left[\dot{A}_i^{(\alpha)}(t)K + \dot{C}_i^{(\alpha)}(t)\right]$$

$$= \frac{\left(a_i e^{\varpi_i t}\right)^2}{4}K + e^{\varsigma_i s}g_i K + \xi_i \frac{\left(b_i e^{\omega_i t}\right)^2}{4} - \frac{[A_i(t)]^2}{4c_i e^{2\kappa_i t}}$$

$$+ \left[\sum_{j=1}^{n} \frac{A_i(t)A_j(t)}{2c_j e^{\kappa_j t}} - A_i(t)\delta K\right], \text{ for } i \in N. \tag{A.1}$$

For (A.1) to hold it is required that

$$\dot{A}_i(t) = (r + \delta)A_i(t) - \frac{\left(a_i e^{\varpi_i t}\right)^2}{4} - e^{\varsigma_i s}g_i, \ A_i(T) = m_i^1; \text{ and}$$

$$\dot{C}_i(t) = rC_i(t) + \frac{[A_i(t)]^2}{4c_i e^{2\kappa_i t}} - \left[\sum_{j=1}^{n} \frac{A_i(t)A_j(t)}{2c_j e^{\kappa_j t}}\right] - \xi_i \frac{\left(b_i e^{\omega_i t}\right)^2}{4},$$

$$C_i(T) = m_i^2, \text{ for } i \in N. \tag{A.2}$$

B.1 Proof of Proposition 4.1

Using Proposition 4.1 and the optimal control strategies in (4.4) the dynamic programming equation in (4.2) can be expressed as:

$$
r\left[A^{(\alpha)}(t)K + C^{(\alpha)}(t)\right] - \left[\dot{A}^{(\alpha)}(t)K + \dot{C}^{(\alpha)}(t)\right]
$$

$$
= \sum_{j=1}^{n} \alpha_j \left[\frac{\left(a_j e^{\varpi_j t}\right)^2}{4} K + e^{\varsigma_j s} g_j K + \xi_j \frac{\left(b_j e^{\omega_i s}\right)^2}{4} - \frac{\left[A^{(\alpha)}(t)\right]^2}{4\alpha_j^2 c_j e^{2\kappa_j t}} \right]
$$

$$
+ A^{(\alpha)} \left[\sum_{j=1}^{n} \frac{A^{(\alpha)}(t)}{2c_j e^{\kappa_j t}} - \delta K \right]. \tag{B.1}
$$

For (B.1) to hold it is required that

$$
\dot{A}^{(\alpha)}(t) = (r+\delta) A^{(\alpha)}(t) - \sum_{j=1}^{n} \alpha_j \left[\frac{\left(a_j e^{\varpi_j t}\right)^2}{4} + e^{\varsigma_j s} g_j \right], \; A^{(\alpha)}(T) = \sum_{j=1}^{n} \alpha_j m_j^1; \text{ and}
$$

$$
\dot{C}^{(\alpha)}(t) = r C^{(\alpha)}(t) + \sum_{j=1}^{n} \frac{\left[A^{(\alpha)}(t)\right]^2}{4\alpha_j c_j e^{2\kappa_j t}} - \sum_{j=1}^{n} \frac{\left[A^{(\alpha)}(t)\right]^2}{2\alpha_j c_j e^{\kappa_j t}} - \sum_{j=1}^{n} \frac{\alpha_j \xi_j \left(b_j e^{\omega_i s}\right)^2}{4},
$$

$$
C^{(\alpha)}(T) = \sum_{j=1}^{n} \alpha_j m_j^2.
$$

$$
\dot{C}^{(\alpha)}(t) = r C^{(\alpha)}(t) - \sum_{j=1}^{n} \frac{\left[A^{(\alpha)}(t)\right]^2}{2\alpha_j c_j e^{\kappa_j t}} - \sum_{j=1}^{n} \frac{\alpha_j \xi_j \left(b_j e^{\omega_i s}\right)^2}{4},
$$

$$
C^{(\alpha)}(T) = \sum_{j=1}^{n} \alpha_j m_j^2. \tag{B.2}
$$

C.1 Proof of Proposition 4.3

Using Proposition 4.3 and the optimal control strategies in (4.4) the dynamic programming equation in (4.9) in Proposition 4.2 can be expressed as:

$$r\left[A_i^{(\alpha)}(t)K + C_i^{(\alpha)}(t)\right] - \left[\dot{A}_i^{(\alpha)}(t)K + \dot{C}_i^{(\alpha)}(t)\right]$$

$$= \left[\frac{\left(a_i e^{\varpi_i t}\right)^2 K}{4} + e^{\varsigma_i s} g_i K + \xi_i \frac{\left(b_i e^{\omega_i t}\right)^2}{4} - \frac{A_i^{(\alpha)}(t,K)}{4c_i e^{2\kappa_i t}(\alpha_i)^2}\right]$$

$$+A_i^{(\alpha)}\left[\sum_{j=1}^{n} \frac{A^{(\alpha)}(t)}{2c_j e^{\kappa_j t}} - \delta K\right], \text{ for } i \in N \text{ and } t \in [0, T]. \qquad (C.1)$$

For (C.1) to hold it is required that

$$\dot{A}_i^{(\alpha)}(t) = (r + \delta) A_i^{(\alpha)}(t) - \frac{\left(a_i e^{\varpi_i t}\right)^2}{4} - e^{\varsigma_i t} g_i, \ A_i^{(\alpha)}(T) = m_i^1; \text{ and}$$

$$\dot{C}_i^{(\alpha)}(t) = rC_i^{(\alpha)}(t) + \frac{A_i^{(\alpha)}(t,K)}{4c_i e^{2\kappa_i t}(\alpha_i)^2} - A_i^{(\alpha)} \sum_{j=1}^{n} \frac{A^{(\alpha)}(t)}{2\alpha_j c_j e^{\kappa_j t}} - \xi_i \frac{\left(b_i e^{\omega_i t}\right)^2}{4}.$$

$$(C.2)$$

References

1. Başar, T., Olsder, G.J. (1995) Dynamic noncooperative game theory. Academic Press, London/New York.
2. Chamberlin, J. (1974) Provision of collective goods as a function of group size. American Political Science Review, 65, 707–716.
3. Dockner, E., Jørgensen, S. (1984) Cooperative and non-cooperative differential games solutions to an investment and pricing problem. The Journal of the International Operational Research Society, 35, 731–739.
4. Dockner, E., Jorgensen, S., Long, N.V., Sorger, G. (2000) Differential games in economics and management science. Cambridge University Press, Cambridge.
5. Fershtman, C., Nitzan, S. (1991) Dynamic voluntary provision of public goods. European Economic Review, 35, 1057–1067.
6. Gradstein, M., Nitzan, S. (1989) Binary participation and incremental provision of public goods. Social Choice and Welfare, 7, 171–192.
7. Hamalainen, R., Haurie, A., Kaitala, V. (1985) Equilibria and threats in a fishery management game. Optimal Control Applications and Methods, 6, 315–333.
8. Leitmann, G. (1974) Cooperative and non-cooperative many players differential games, Springer Verlag, New York.

9. Marin-Solano, J. (2014) Time-consistent equilibria in a differential game model with time inconsistent preferences and partial cooperation. Forthcoming in J. Haunschmied, V. Veliov and S. Wrzaczek (eds.) Dynamic games in economics, Springer Verlag, New York.
10. McGuire, M. (1974) Group size, group homogeneity, and the aggregate provision of a pure public good under Cournot behavior. Public Choice, 18, 107–126.
11. Sorger, G. (2006) Recursive bargaining over a productive asset. Journal of Economic Dynamics and Control, 30, 2637–2659.
12. Wang, W.-K., Ewald, C.-O. (2010) Dynamic voluntary provision of public goods with uncertainty: a stochastic differential game model. Decisions in Economics and Finance, 3, 97–116.
13. Wirl, F. (1996) Dynamic voluntary provision of public goods: Extension to nonlinear strategies. European Journal of Political Economy, 12, 555–560.
14. Yeung, D.W.K. (2004) Nontransferable individual payoff functions under stochastic dynamic cooperation, International Game Theory Review, 6, 281–289.
15. Yeung, D.W.K., Petrosyan, L.A. (2005) Subgame consistent solution of a cooperative stochastic differential game with nontransferable payoffs, Journal of Optimization Theory Application, 124(3), 701–724.
16. Yeung, D.W.K., Petrosyan, L.A. (2012) Subgame consistent economic optimization: An advanced cooperative dynamic game analysis. Birkhäuser, Boston.
17. Yeung, D.W.K., Petrosyan, L.A. (2013) Subgame consistent cooperative provision of public goods. Dynamic Games and Applications. https://doi.org/10.1007/s13235-012-0062-7.
18. Yeung, D.W.K., Petrosyan, L.A. (2014) Subgame consistent cooperative provision of public goods under accumulation and payoff uncertainties. In J. Haunschmied, V.M. Veliov and S. Wrzaczek (eds), Dynamic games in economics, 289–315. Springer-Verlag Berlin Heidelberg.
19. Yeung, D.W.K., Petrosyan, L.A. (2015) Subgame consistent cooperative solution for NTU dynamic games via variable weights, Automatica, 50, 84–89.
20. Yeung, D.W.K., Petrosyan, L.A. (2016) Subgame consistent cooperation: A comprehensive treatise. Springer, Singapore.
21. Yeung, D.W.K., Petrosyan, L.A., Yeung, P.M. (2007) Subgame consistent solutions for a class of cooperative stochastic differential games with nontransferable payoffs. Annals of the International Society of Dynamic Games, 9, 153–170.

Strongly Time-Consistent Solutions in Cooperative Dynamic Games

Leon A. Petrosyan

Abstract In the paper the evolution of dynamic game along the cooperative trajectory is investigated. Along cooperative trajectory at each time instant players find themselves in a new game which is a subgame of the originally defined game. In many cases the optimal solution of the initial game restricted to the subgame along cooperative trajectory fails to be optimal in the subgame. To overcome this difficulty we introduced (see Petrosyan and Danilov, Vestnik Leningrad Univ Mat Mekh Astronom 1:52–59, 1979; Petrosyan and Zaccour, J Econ Control 27(3):381–398, 2003; Yeung and Petrosyan, Subgame consistent economic optimization. Birkhauser, 2012) the special payment mechanism—imputation distribution procedure (IDP), or payment distribution procedure (PDP), but another serious question arises: under what conditions the initial optimal solution converted to any optimal solution in the subgame will remain optimal in the whole game. This condition we call strongly time-consistency condition of the optimal solution. If this condition is not satisfied players in reality may switch in some time instant from the previously selected optimal solution to any optimal solution in the subgame, and as result realize the solution which will be not optimal in the whole game. We propose different types of strongly time-consistent solutions for multicriterial control, cooperative differential, and cooperative dynamic games.

Keywords Differential game · Time consistency · Dynamic stability · Pareto optimality · Cooperation

1 What Is Strongly Time-Consistency?

What is strongly time-consistency? Try to explain this notion. Let $M \in R^n$ be a fixed point in R^n. Consider a classical control problem (with one player)

L. A. Petrosyan (✉)
St. Petersburg State University, Saint Petersburg, Russia
e-mail: l.petrosyan@spbu.ru

© Springer Nature Switzerland AG 2020
D. Yeung et al. (eds.), *Frontiers in Games and Dynamic Games*, Annals of the
International Society of Dynamic Games 16,
https://doi.org/10.1007/978-3-030-39789-0_2

$$\dot{x} = f(x, u), x \in R^n, u \in U \subset Comp\, R^l$$
$$x(t_0) = x_0, \ t \in [t_0, T].$$

(1)

Find the control $\bar{u}(t)$, and corresponding trajectory $\bar{x}(t)$ such that at terminal instant the distance $\rho(\bar{x}(T), M)$ will be minimal.

Denote this problem by $\Gamma(x_0, T - t_0)$. And denote by $C(x_0, T - t_0)$ the reachability set of system (1) from initial point x_0 at terminal time T.

Suppose for simplicity that $M \notin C(x_0, T - t_0)$. The solution of this optimal control problem we can see on Fig. 1.

Consider the intermediate time instant $\tau \in [t_0, T]$, and the intermediate control problem $\Gamma(\bar{x}(\tau), T - \tau)$ with initial condition on the optimal trajectory with duration $T - \tau$. It is clear that the control $\bar{u}(t), t \in [\tau, T]$ will be optimal also in $\Gamma(\bar{x}(\tau), T - \tau)$, so will be also the trajectory $\bar{x}(t), t \in [\tau, T]$.

This is Bellman-optimality principle and also time-consistency of optimal control $\bar{u}(t), t \in [t_0, T]$. Suppose now that we have another optimal control $\bar{\bar{u}}(t), t \in [\tau, T]$ in the problem $\Gamma(\bar{x}(\tau), T - \tau)$. Then it is easy to see that the control

$$\hat{u}(t) = \begin{cases} \bar{u}(t), \ t \in [t_0, \tau] \\ \bar{\bar{u}}(t), \ t \in [\tau, T] \end{cases}$$

will be also optimal in the problem $\Gamma(x_0, T - t_0)$. In other words: "any optimal continuation of the original problem in the subproblem along optimal trajectory generates optimal solution of the original problem." This property we shall call *strongly time-consistency (strongly dynamic stability)* (see Fig. 1).

Fig. 1 Classical optimal control problem

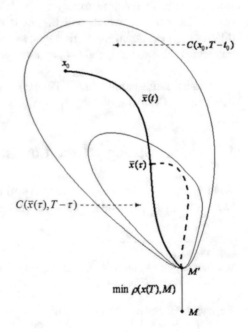

Consider now a slightly more complicated problem. The motion equations are the same (1), but the aim of control is different, it is necessary to come as close as possible to system of points $M_1, \ldots, M_k$, $M_i \in R^n$, $i \in \{1, \ldots k\}$.

Denote as before the problem by $\Gamma(x_0, T - t_0)$ and by $C(x_0, T - t_0)$ the reachability set of (1) and suppose that $C(x_0, T - t_0) \cap \hat{M} = \emptyset$, where $\hat{M}$ is the convex hull of points $\{M_1, \ldots, M_k\}$. As optimal solution here we may consider Pareto-optimal set which coincides with arc AB, the projection (suppose that $C(x_0, T - t_0)$ is convex) of $\hat{M}$ on $C(x_0, T - t_0)$ (see Fig. 2).

Consider Pareto-optimal control $\bar{u}(t)$, $t \in [t_0, T]$ which connects the initial point $x_0 \in C(x_0, T - t_0)$ with the point M belonging to the Pareto-optimal set (M belongs to the arc AB which is projection of the set $\hat{M}$ on $C(x_0, T - t_0)$). And let $\bar{x}(t)$, $t \in [t_0, T]$ be the corresponding Pareto-optimal trajectory.

Consider a subproblem $\Gamma(\bar{x}(t), T - t)$ from initial position $\bar{x}(t)$ on the Pareto-optimal trajectory. We see that the Pareto-optimal set in $\Gamma(\bar{x}(t), T - t)$ (arc $A'B'$) is different from the Pareto-optimal set in $\Gamma(x_0, T - t_0)$ having only (in our example) one common point M. This means that the control $\bar{u}(t)$, $t \in [\tau, T]$ is Pareto-optimal in subproblem $\Gamma(\bar{x}(\tau), T - \tau)$, and the Pareto-optimal solution $\bar{u}(t)$, $t \in [t_0, T]$ is time-consistent (dynamic stable) [4, 5].

In the same time we can see that the control of the type

$$\hat{u}(t) = \begin{cases} \bar{u}(t), \, t \in [t_0, \tau] \\ \bar{\bar{u}}(t), \, t \in [\tau, T], \end{cases}$$

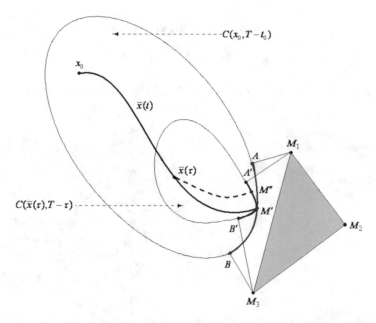

Fig. 2 Multicriterial optimal control problem

where $\bar{\bar{u}}(t)$ is an arbitrary Pareto-optimal control in subproblem $\Gamma(\bar{x}(\tau), T - \tau)$, may not be Pareto-optimal in $\Gamma(x_0, T - t_0)$.

Which means that in this case the optimal continuation of the motion in the subproblem with initial conditions on Pareto-optimal trajectory together with initial Pareto-optimal motion maybe not Pareto-optimal in the original problem. This means that the Pareto-optimal solution is time-consistent but not strongly time-consistent (see Fig. 2).

In this special problem there is one approach for constructing strongly time-consistent solutions on the bases of Pareto-optimal solutions. The idea of this approach is to consider all possible outcomes which may occur if at each time instant t on the time interval $[t_k, t_k + \delta)$ the control $u(\tau)$ will be selected leading to one of Pareto-optimal points in the subproblem $\Gamma(x(t_k), T - t_k)$. Let $t_0 < t_1 < \ldots < t_k < t_{k+1} < \ldots < t_n = T$ be the decomposition of the time interval $[t_0, T]$, $t_{k+1} - t_k = \delta > 0$. The resulting trajectory will be not Pareto-optimal, but we shall call it conditionally Pareto-optimal. Denote by $P(x(t_k), t_k)$ the set of end-points of these trajectories for all possible controls selected in a described manner. It is clear that

$$P(x(t_0), t_0) \supset P(x(t_1), t_1) \supset \ldots \supset P(x(t_k), t_k) \supset \ldots \supset P(x(T), T).$$

And the set $P(x(t_0), t_0)$ is δ-strongly time-consistent if we allow possible changes of controls only in points $t_k, k = 0, \ldots, n$.

For the system

$$\dot{x} = u_1 + u_2 + u_3, x(t_0) = x_0$$

$$|u_i| \leq 1, x \in R^2, t \in [t_0, T],$$

the set $P(x(t_0), t_0)$ is denoted by $\hat{D}$ on the Fig. 3 (dashed region).

Fig. 3 Example of strongly time-consistent solution

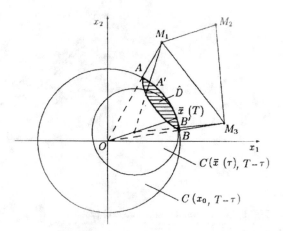

1.1 Cooperative Differential Game

Consider now cooperative differential games with player set N. Motion equations have the form

$$\dot{x} = f(x, u_1, \ldots, u_k), \ x \in R^n, u_i \in U_i \subset Comp R^l \tag{2}$$

$$x(t_0) = x_0 \tag{3}$$

and the payoffs of players are defined as

$$K_i(x_0, T - t_0; u_1, \ldots, u_k) = \int_{t_0}^{T} h_i(x(t))dt, h_i > 0, \ i \in N.$$

Denote this game by $\Gamma(x_0, T - t_0)$. Cooperative trajectory $\bar{x}(t)$, $\bar{x}(t_0) = x_0$, $t \in [t_0, T]$ is defined as

$$\max_{u_1, \ldots, u_k} \sum_{i=1}^{n} K_i(x_0, T - t_0; u_1, \ldots, u_n) = \sum_{i=1}^{n} K_i(x_0, T - t_0; \bar{u}_1, \ldots, \bar{u}_n) =$$

$$= \sum_{i=1}^{n} \int_{t_0}^{T} h_i(\bar{x}(t))dt = v(x_0, T - t_0; N). \tag{4}$$

We suppose that max in (4) is attained. Let $v(x_0, T - t_0; S)$, $S \subset N$ be the characteristic function defined in classical way as value of zero-sum game between coalition S as first player and $N \backslash S$ as second (see [6]), and $E(x_0, T - t_0)$, the set of imputations

$$E(x_0, T - t_0) = \{\xi = \{\xi_i\} : \sum_{i=1}^{n} \xi_i = v(x_0, T - t_0; N),$$

$$\xi_i \geq v(x_0, T - t_0; \{i\}), i \in N\}. \tag{5}$$

Denote by $C(x_0, T - t_0)$ reachability set of the system (1), for $y \in C(x_0, t - t_0)$, $t \in [t_0, T]$ define a subgame $\Gamma(y, T - t)$ of $\Gamma(x_0, T - t_0)$ with characteristic function $v(y, T - t; S)$, $S \subset N$ and imputation set $E(y, T - t)$.

Optimality principle (solution) is a subset of imputation set

$$C(y, T - t) \subset E(y, T - t)$$

(Core, NM-solution,...).

Consider the family of subgames along the cooperative trajectory $\Gamma(\bar{x}(t), T - t; S)$ and also imputation set $E(\bar{x}(t), T - t)$ and the solution of subgames along this cooperative trajectory, $C(\bar{x}(t), T - t)$.

For each $\xi \in C(x_0, T - t_0)$ define the imputation distribution procedure IDP [3] $\beta(t) = (\beta_1(t), \ldots, \beta_i(t), \ldots, \beta_n(t))$

$$\xi = \int_{t_0}^{T} \beta(\tau)d\tau, \ \xi \in C(x_0, T - t_0).$$

The imputation $\xi \in C(x_0, T - t_0)$ is called dynamic stable [3–5] (time-consistent) if

$$\xi - \int_{t_0}^{t} \beta(\tau)d\tau \in C(\bar{x}(t), T - t), \ t \in [t_0, t].$$

Definition 1 The solution $C(x_0, T - t_0)$ is called time-consistent if all imputations $\xi \in C(x_0, T - t_0)$ are time-consistent.

Definition 2 Optimality principle $C(x_0, T - t_0)$ is called strongly dynamic stable [11] (strongly time-consistent) if for each $\xi \in C(x_0, T - t_0)$ there exist IDP $\beta(\tau)$ such that

$$\int_{t_0}^{t} \beta(\tau)d\tau \oplus C(\bar{x}(t), T - t) \subset C(x_0, T - t_0),$$

here $a \oplus B(a \in R^n, B \subset R^n)$ is defined as $\{a + b : b \in B\}$.

Since as it is well known time-consistency of cooperative solutions taken from the classical one-shot game theory takes place only in special cases it is clear that strongly time-consistency is a very special event. Note that strongly time-consistency has sense only for multivalued (set-valued) optimality principles (core, NM-solution).

1.2   Transformation of Characteristic Function

Let $v(y, T - t; S)$ be characteristic function in $\Gamma(y, T - t)$. Define the following integral transformation

$$\bar{v}(x_0, T - t_0; S) = \int_{t_0}^{T} \frac{v(\bar{x}(t), T - t; S) \sum_{i \in N} h_i(\bar{x}(t))}{v(\bar{x}(t), T - t; N)} dt,$$

here $v(\bar{x}(t), T - t; S)$ is characteristic function computed for subgame $\Gamma(\bar{x}(t), T - t)$ along cooperative trajectory. It can be seen that

$$\overline{v}(x_0, T - t; N) = v(x_0, T - t; N).$$

Define the imputation set $\overline{E}(x_0, T - t_0)$ and the core under the new characteristic function $\overline{v}(x_0, T - t_0; S)$, $\overline{C}(x_0, T - t_0) \subset \overline{E}(x_0, T - t_0)$ and define the integral transformation of the imputation $\xi \in E(x_0, T - t_0)$ to $\overline{\xi} \in \overline{E}(x_0, T - t_0)$ as

$$\overline{\xi}_i = \int_{t_0}^{T} \frac{\xi_i(t) \sum\limits_{i \in N} h_i(\overline{x}(t))}{V(\overline{x}(t), T - t; N)} dt, \ i \in N,$$

where $\xi(t) \in E(\overline{x}(t), T - t)$. Similarly let $\overline{E}(\overline{x}(t), T - t)$ $\overline{C}(\overline{x}(t), T - t)$ be the set of imputations and the core in subgame $\Gamma(\overline{x}(t), T - t)$ along cooperative trajectory under characteristic function

$$\overline{v}(\overline{x}(t), T - t; S) = \int_{t}^{T} \frac{v(\overline{x}(\tau), T - \tau; S) \sum\limits_{i \in N} h_i(\overline{x}(\tau))}{v(\overline{x}(\tau), T - \tau; N)} d\tau.$$

Theorem 1 $\overline{C}(x_0, T - t_0)$ *is strongly time-consistent.*

To prove it is sufficient to take for each $\overline{\xi} \in \overline{E}(x_0, T - t_0)$ as $\beta_i(t)$

$$\beta_i(t) = \frac{\xi_i(t) \sum\limits_{i \in N} h_i(\overline{x}(t))}{v(\overline{x}(t), T - t; N)},$$

where $\xi(t) \in C(\overline{x}(t), T - t)$ is an integrable selector from $C(\overline{x}(t), T - t)$.

What is the connection between $\overline{C}$ and C? If there is a nonvoid intersection of $\overline{C}$ and C, then this imputation set could be a good preferable optimality principle in $\Gamma(x_0, T - t)$. Introduce

$$\lambda(S) = \max_{t_0 \leq t \leq T} \frac{v(\overline{x}(t), T - t; S)}{v(\overline{x}(t), T - t; N)},$$

$$\lambda(N) = 1.$$

We have

$$\overline{v}(x_0, T - t_0; S) \leq \lambda(S) \int_{t_0}^{T} \sum_{i \in N} h_i(\overline{x}(t)) dt = \lambda(S) v(x_0, T - t_0; N),$$

$$\overline{v}(x_0, T - t_0; N) = \lambda(N) v(x_0, T - t_0; N) = v(x_0, T - t_0; N),$$

$$\lambda(S) \geq \frac{v(x_0, T - t_0; S)}{v(x_0, T - t_0; N)},$$

$$v(x_0, T - t_0; S) \leq \lambda(S) v(x_0, T - t_0; N).$$

Denote by $\hat{C}(x_0, T - t_0)$ the set of all solutions $\xi = \{\xi_1, \ldots, \xi_n\}$

$$\sum_{i \in S} \xi_i \geq \lambda(S)v(x_0, T - t_0; N), \quad S \subset N, \quad \sum_{i \in N} \xi_i = v(x_0, T - t_0; N).$$

From previous considerations it follows

$$\sum_{i \in S} \xi_i \geq \lambda(S)v(x_0, T - t_0; N) \geq v(x_0, T - t_0; S).$$

We see that

$$\hat{C}(x_0, T - t_0) \subset C(x_0, T - t_0) \cap \overline{C}(x_0, T - t_0)$$

and

$$\hat{C}(\bar{x}(t), T - t) \subset C(\bar{x}(t), T - t) \cap \overline{C}(\bar{x}(t), T - t).$$

The following theorem holds.

Theorem 2

$$\overline{C}(x_0, T - t_0) \supset \int_{t_0}^{t} \frac{\xi(t) \sum_{i=1}^{n} h_i(\bar{x}(t))}{v(\bar{x}(t), T - t; N)} \oplus \hat{C}(\bar{x}(t), T - t) \tag{6}$$

for any integrable selector $\xi(t) \in C(x(t), T - t)$.

Proof Theorem 2 follows from the inclusion $\hat{C}(\bar{x}(t), T - t) \subset \overline{C}(\bar{x}(t), T - t)$ and strongly time-consistency of $\overline{C}(x_0, T - t_0)$.

From Theorem 2 it follows that for each imputation $\xi_0 \in C(x_0, T - t_0) \cap \hat{C}(x_0, T - t_0)$ there exist IDP

$$\beta(t) = \frac{\xi(t) \sum_{i=1}^{n} h_i(\bar{x}(t))}{v(\bar{x}(t), T - t; N)},$$

where $\xi(t_0) = \xi_0$ and $\xi(t)$ is an integrable selector from $C(\bar{x}(t), T - t)$, such that

$$\int_{t_0}^{t} \beta(\tau)d\tau \oplus \hat{C}(\bar{x}(t), T - t) \subset \overline{C}(x_0, T - t_0). \tag{7}$$

$\square$

Suppose that $\hat{C}(x_0, T - t_0) \neq \emptyset$. The interpretation of (7) is as follows. $\hat{C}(x_0, T - t_0)$ is the subset of the original core $C(x_0, T - t_0)$ and for any imputation $\xi \in \hat{C}(x_0, T - t_0) \cap C(x_0, T - t_0)$ from this subset of original core $C(x_0, T - t_0)$ one can construct the IDP (the imputation distribution procedure) such that if in an intermediate time instant t players for some reasons would like to switch to another optimal imputation $(\xi^t)' \in \hat{C}(\bar{x}(t), T - t) \subset C(\bar{x}(t), T - t)$ from the subset of original core, they will still get the payments according to the imputation from $\bar{C}(x_0, T - t_0)$, resulting from the integral transformation of $C(x_0, T - t_0)$.

2 Repeated Games

Folk theorems are well known in game theory [1, 2, 6–9]. By using the so-called punishment strategies they show the possibility to attain in some sense preferable outcomes. These outcomes are stable against deviations of single players. But the natural question arises: is it possible to get "good" outcomes stable against deviations of coalitions (coalition-proofness). Now we try to construct a mechanism based on the introduction of an analog of characteristic function which makes it possible (under some conditions on this newly defined characteristic function) to get coalition-proofness for repeated and multistage games [9]. This will show us the way of constructing strongly time-consistent optimality principles in multistage games.

Denote by G the infinity repeated n-person game with the game Γ played on each stage. For simplicity suppose that the stage game Γ is finite (has finite sets of strategies).

$$\Gamma = < N; U_1, \ldots, U_i, \ldots, U_n; K_1, \ldots, K_i, \ldots, K_n > .$$

If on stage $k(1 \leq k \leq \infty)$ strategy profile $u^k = (u_1^k, \ldots, u_i^k, \ldots, u_n^k)$ is chosen, the payoff in G is defined as

$$
\begin{aligned}
H_i(u_1(\cdot), \ldots, u_i(\cdot), \ldots, u_n(\cdot)) &= \sum_{k=1}^{\infty} \delta^{k-1} K_i(u_1^k, \ldots, u_i^k, \ldots, u_n^k) = \\
&= \sum_{k=1}^{\infty} \delta^{k-1} K_i(u^k) = H_i(u(\cdot)), \ i \in N,
\end{aligned}
\tag{8}
$$

here $u_1(\cdot) = (u_1^1, \ldots, u_1^k, \ldots), \ldots, u_i(\cdot) = (u_i^1, \ldots, u_i^k, \ldots), \ldots, u_n(\cdot) = (u_n^1, \ldots, u_n^k, \ldots), \delta \in (0, 1)$.

Here in the expression $u_i(\cdot) = (u_i^1, \ldots, u_i^k, \ldots), i \in N$ u_i^k is the strategy chosen by player i in the game Γ on stage k. We suppose that on stage k when choosing u_i^k player i knows the choices of other players and remembers his choices on previous stages. Thus u_i^k is function of history

$$h^k = (u_1^1, \ldots, u_1^{k-1}; \ldots; u_i^1, \ldots, u_i^{k-1}; \ldots; u_n^1, \ldots, u_n^{k-1}).$$

Formally we have to write $u_i^k(h^k)$, i.e. u_i^k depends upon history h^k, $k = 1, \ldots$. However in this paper for convenience we shall write u_i^k instead $u_i^k(h^k)$.

Consider the strategy profile $\bar{u}(\cdot) = (\bar{u}_1(\cdot), \ldots, \bar{u}_i(\cdot), \ldots, \bar{u}_n(\cdot))$ such that

$$\sum_{i \in N} H_i(\bar{u}) = \max_{u(\cdot)} \sum_{i \in N} H_i(u). \tag{9}$$

It is evident that such strategy profile always exists.

One can take $\bar{u}_i(\cdot) = (\bar{u}_i^1, \ldots, \bar{u}_i^k, \ldots,) \, i \in N$ such that

$$\sum_{i \in N} K_i(\bar{u}_1, \ldots, \bar{u}_i, \ldots, \bar{u}_n) = \max_{u_1, \ldots, u_i, \ldots, u_n} \sum_{i \in N} K_i(u_1, \ldots, u_i, \ldots, u_n) \tag{10}$$

and since the stage games are the same (G is repeated game) we can take $\bar{u}_i^k = \bar{u}_i$ for all $k = 1, \ldots, n$. Then from (8)–(10) we get that

$$\sum_{i \in N} H_i(\bar{u}) = \sum_{i \in N} \left(\sum_{k=1}^{\infty} \delta^{k-1} K_i(\bar{u}_1^k, \ldots, \bar{u}_n^k) \right) =$$

$$= \sum_{i \in N} \left(\sum_{k=1}^{\infty} \delta^{k-1} K_i(\bar{u}_1, \ldots, \bar{u}_n) \right) = \frac{1}{1 - \delta} \sum_{i \in N} K_i(\bar{u}_1, \ldots, \bar{u}_n). \tag{11}$$

Introduce characteristic function $V(S)$, $S \subset N$ in Γ in classical sense. Then we shall have

$$V(N) = \sum_{i \in N} K_i(\bar{u}_1, \ldots, \bar{u}_n) \tag{12}$$

and it can be easily shown that the characteristic function $W(S)$, $S \subset N$ in G will have the form

$$W(S) = \frac{1}{1 - \delta} V(S), \; S \subset N. \tag{13}$$

Remind now the definition of strong (or coalition proof) Nash equilibrium.

Definition 3 The n-tuple of strategies $(\hat{u}_1, \ldots \hat{u}_2, \ldots \hat{u}_n) = \hat{u}$ is called strong (or coalition proof) Nash equilibrium (SNE) if for all $S \subset N$, and all $u_S = \{u_i, i \in S\}$ the following inequality holds

$$\sum_{i \in S} K_i(\hat{u}) \geq \sum_{i \in S} K_i(\hat{u} \| u_S). \tag{14}$$

Consider now the core C in Γ, and suppose that $C \neq \emptyset$, and suppose also that there exist an imputation $\alpha \in C$ such that

$$\sum_{i \in S} \alpha_i > V(S), \ S \subset N, \ S \neq N. \tag{15}$$

2.1 Associated Zero-Sum Games

Consider a family of zero-sum games $\Gamma_{N \backslash i, i}$ with coalition $N \backslash \{i\}$ as first player and coalition $\{i\}$ as second. The payoff of $N \backslash \{i\}$ is equal to the sum of payoffs of players from $N \backslash \{i\}$. Denote by $V(N \backslash i)$ the value of $\Gamma_{N \backslash i, i}$. Let $(\bar{\mu}_{N \backslash i}, \bar{\mu}_i)$ be the saddle point (in mixed strategies) in $\Gamma_{N \backslash i, i}$.

Consider the n-tuple of strategies $\bar{\mu} = (\bar{\mu}_1, \ldots, \bar{\mu}_n)$, and define

$$\overline{W}(S) = \max_{\mu_S} \sum_{i \in S} K_i(\mu_S; \bar{\mu}_{N \backslash S}),$$

here $\mu_S = \{\mu_i, \ i \in S\}, \bar{\mu}_{N \backslash S} = \{\bar{\mu}_i, \ i \in N \backslash S\}$. It is clear that

$$\overline{W}(S) \geq V(S), \ \overline{W}(N) = V(N), \ S \subset N.$$

Suppose, that there exist the solution of the system

$$\sum_{i \in S} \alpha_i > \overline{W}(S), \ \sum_{i \in N} \alpha_i = \overline{W}(N) = V(N). \tag{16}$$

Construct now the modification G^α of the game G. The difference between G^α and G is in payoffs defined in stage games Γ when the cooperative strategies $\bar{u} = (\bar{u}_1, \ldots, \bar{u}_n)$ are used and the payoff in this case is equal to $\alpha = (\alpha_1, \ldots, \alpha_n)$, where α satisfies (16). For all other strategy combinations the payoffs remain as in Γ.

The following theorem holds [10].

Theorem 3 *In game G^α there exist $\delta \in (0, 1)$ and SNE such that payoffs in this SNE are equal to $\alpha_i \dfrac{1}{1 - \delta}$, which are payoffs in G^α under cooperation.*

2.2 Multistage Games

Multistage game G starts from a fixed stage game $\Gamma(z_1)$ which can be considered as situated in the position (root) z_1 of the game tree G.

$$\Gamma(z_1) = < N; U_1^{z_1}, \ldots, U_i^{z_1}, \ldots, U_n^{z_1}; K_1^{z_1}, \ldots, K_i^{z_1}, \ldots, K_n^{z_1} > . \tag{17}$$

For simplicity we suppose that the set of players N is the same in all stage games. When the game G develops the infinite sequence of stage games is realized but only a finite number of them are different since we suppose that the total number of different stage game $\Gamma(z)$ is finite. As usual in multistage games we consider the general case when the next stage game depends upon controls chosen by players only in previous stage game. Like in previous section denote by $u_i(\cdot)$ the strategy of player i in G (defined as function of histories). The strategy profile which maximizes the sum of players payoffs in G is called "cooperative" strategy profile and the corresponding sequence of stage games (or equivalently sequence of positions on the tree G) "cooperative trajectory." Suppose that for each stage game $\Gamma(z)$ the characteristic function $V(z, S)$ (in classical sense) is defined.

For each stage game $\Gamma(z)$ consider the family of zero-sum games $\Gamma_{N\setminus i,i}(z)$ and corresponding saddle points $\bar{\mu}^z_{N\setminus i}$, μ^z_i, and $\bar{\mu}^z = (\bar{\mu}^z_1, \ldots, \bar{\mu}^z_n)$, define

$$\overline{W}(z, S) = \max_{\mu^z_S} \sum_{i \in S} K^z_i(\mu^z_S, \bar{\mu}^z_{N\setminus S}).$$

Let

$$\overline{W}(S) = \sup_z \overline{W}(z, S).$$

Suppose that

$$\overline{W}(S) < \inf_z W(z, N) = \inf_z V(z, N).$$

Suppose the core $C(z)$ is not empty in each stage game $\Gamma(z)$, denote by $D(z)$ the subcore of $C(z)$ as set of all imputations $\alpha^z = (\alpha^z_1, \ldots, \alpha^z_n)$, $\sum_{i \in S} \alpha^z_i \geq \overline{W}(S)$, for all S.

Suppose that for all $z \in G$, $D(z) \neq \emptyset$ and suppose also that there exist imputation $\alpha^z = (\alpha^z_1, \ldots, \alpha^z_n)$ such that

$$\sum_{i \in S} \alpha^z_i > \overline{W}(S) \text{ for all } S, \tag{18}$$

$$\inf_{S,z} \left[\sum_{i \in S} \alpha^z_i - \overline{W}(S) \right] = A > 0. \tag{19}$$

For simplicity we shall consider the special case when $V(z, N) = \overline{W}(N)$ for all z the previous conditions (18) and (19) can be written as

$$\sum_{i \in S} \alpha_i > \overline{W}(S) \text{ for all } S, \tag{20}$$

$$\inf_S \left[\sum_{i \in S} \alpha_i - \overline{W}(S) \right] = A > 0, \tag{21}$$

since the number of different stage games is finite and we can select α the same in all stage games.

Construct now the modification G^α of the game in the same way as it was done in Sect. 1. Theorem 1 from Sect. 1 holds also for the game G^α.

Theorem 4 *In the game G^α there exist $\delta \in (0, 1)$ and SNE such that payoffs in this SNE are equal to $\alpha_i \frac{1}{1-\delta}$, which are payoffs in G^α under cooperation.*

2.3 Time-Consistency and Strongly Time-Consistency

Consider cooperative version of game G and subgame $G(z)$. Introduce the following characteristic function in G and in $G(z)$, respectively,

$$\hat{W}(S) = \frac{1}{1-\delta} W(S).$$

Denote the analog of the core $\hat{C}$ and $\hat{C}(z)$ in G under the defined above c.f.

Strongly time-consistency in this case means that for each imputation $\bar{\alpha} \in \hat{C}(\bar{z}_0)$ there exist corresponding IDP $\bar{\beta}(1), \ldots, \bar{\beta}(l), \ldots$ such that

$$\sum_{k=0}^{l} \delta^k \bar{\beta}(k) \oplus \delta^{l+1} \hat{C}(\bar{z}_{l+1}) \subset \hat{C}(\bar{z}_0). \tag{22}$$

It can be easily seen that if $D(z) = D \neq \emptyset$, by selecting $\bar{\beta}(k) = \beta \in D(\bar{z}_k)$ we can guarantee the strongly time-consistency of $\hat{C}(\bar{z}_0)$.

Suppose $\alpha \in \hat{C}(\bar{z}_0)$, then by definition we have

$$\sum_{i \in S} \bar{\alpha}_i \geq \hat{W}(S) = \frac{1}{1-\delta} \overline{W}(S); \quad \sum_{i \in N} \bar{\alpha}_i = \hat{W}(N) = \frac{1}{1-\delta} \overline{W}(N).$$

Represent $\bar{\alpha}$ in the form

$$\bar{\alpha} = \sum_{k=0}^{\infty} \delta^k \bar{\beta},$$

since $\bar{\alpha} \in \hat{C}(\bar{z}_0)$

$$\sum_{i \in S} \bar{\alpha}_i = \sum_{i \in S} \frac{1}{1-\delta} \bar{\beta}_i \geq \hat{W}(S) = \sum_{i \in S} \frac{1}{1-\delta} \bar{W}(S),$$

and

$$\sum_{i \in S} \bar{\beta}_i \geq \bar{W}(S), \quad \sum_{i \in N} \bar{\beta}_i = \bar{W}(N).$$

Thus $\bar{\beta} \in D(\bar{z}_k) = D$, $k = 0, 1, \ldots, l, \ldots$. And we get that each imputation $\bar{\alpha} \in \hat{C}(\bar{z}_0)$ can be represented in the form $\bar{\alpha} = \sum_{k=0}^{\infty} \delta^k \bar{\beta}(k)$, when $\bar{\beta}(k) = \bar{\beta} \in D(\bar{z}_k) = D$.

This will give us also strongly time-consistency of $\hat{C}(\bar{z}_0)$.

We have seen that for arbitrary $\bar{\alpha} \in \hat{C}(\bar{z}_0)$ there exist such IDP $\bar{\beta}(0), \bar{\beta}(1), \ldots,$ $\bar{\beta}(k), \ldots$ (in our case $\bar{\beta}(k) = \bar{\beta} \in D$), that

$$\bar{\alpha} = \sum_{k=0}^{\infty} \delta^k \bar{\beta}(k).$$

Suppose that $\alpha' \in \sum_{k=0}^{l} \delta^k \bar{\beta}(k) \oplus \delta^{l+1} \hat{C}(\bar{z}_{l+1})$. To prove (22) we have to prove that in this case $\alpha' \in \hat{C}(\bar{z}_0)$. Consider the stage l then we can write the imputation α' in the form

$$\alpha' = \sum_{k=0}^{l} \delta^k \bar{\beta}(k) + \delta^{l+1} \alpha'',$$

here $\bar{\beta}(k) = \bar{\beta} \in D$, where $\alpha'' \in \hat{C}(\bar{z}_{l+1})$.

Since $\alpha'' \in \hat{C}(\bar{z}_{l+1})$ we have

$$\sum_{i \in S} \alpha_i'' \geq \hat{W}(S) = \frac{1}{1-\delta} \bar{W}(S), \quad \sum_{i \in N} \alpha_i'' = \hat{W}(N) = \frac{1}{1-\delta} \bar{W}(N),$$

and we can show that similar to previous case when $\alpha \in \hat{C}(\bar{z}_0)$, α'' can be represented in the form

$$\alpha'' = \sum_{k=l+1}^{\infty} \delta^{k-(l+1)} \beta''(k),$$

where $\beta''(k) = \beta'' \in D$, $k = l + 1, \ldots$.

Then we get

$$\alpha' = \sum_{k=0}^{l} \delta^k \bar{\beta}(k) + \delta^{l+1} \sum_{k=l+1}^{\infty} \delta^{k-(l+1)} \bar{\bar{\beta}}(k) = \sum_{k=0}^{\infty} \delta^k \tilde{\beta}(k),$$

where $\tilde{\beta}(k) \in D, \tilde{\beta}(k) = \bar{\beta}(k) = \bar{\beta}, k = 1, \dots, l, \tilde{\beta}(k) = \bar{\bar{\beta}}(k) = \beta'', k = l+1, \dots.$
And we have

$$\sum_{i \in S} \alpha' = \sum_{k=0}^{l} \delta^k \sum_{i \in S} \tilde{\beta}_i(k) + \sum_{k=l+1}^{\infty} \delta^k \sum_{i \in S} \tilde{\beta}_i(k) = \sum_{k=1}^{l} \delta^k \sum_{i \in S} \bar{\beta}_i(k) + \sum_{k=l+1}^{\infty} \delta^k \sum_{i \in S} \bar{\bar{\beta}}_i(k) \geq,$$

$$\geq \sum_{k=0}^{l} \delta^k \bar{W}(S) + \sum_{k=l+1}^{\infty} \delta^k \bar{W}(S) = \sum_{k=0}^{\infty} \delta^k \bar{W}(S) = \frac{1}{1-\delta} \bar{W}(S) = \hat{W}(S).$$

In the similar way we can prove that $\sum_{i \in N} \alpha'_i = \hat{W}(S)$. This proves that $\alpha' \in \hat{C}(\bar{z}_0)$.

Acknowledgement This research was supported by the Russian Science Foundation (grant 17-11-01079).

References

1. Aumann, R.J., Maschler, M.: Repeated Games with Incomplete Information. MIT Press, Cambridge (1995)
2. Myerson, R.B.: Multistage Games with Communication. Econometrica. **54**, 323–358 (1986)
3. Petrosyan, L.A., Danilov, N.N.: Stability of the solutions in nonantagonistic differential games with transferable payoffs. Vestnik Leningrad. Univ. Mat. Mekh. Astronom. **1**, 52–59 (1979)
4. Petrosyan, L.A., Zaccour, G.: Time-consistent Shapley value allocation of pollution cost reduction. Journal of Economics and Control. **27**, 3, 381–398 (2003)
5. Yeung, D.W.K., Petrosyan, L.A.: Subgame Consistent Economic Optimization. Birkhauser (2012)
6. M. Maschler, M., Solan, E., Zamir, S.: Game Theory. Cambridge University Press (2013)
7. Aumann, R., Shapley, L.: Long-Term Competition – A Game-Theoretic Analysis. Essays in Game Theory. (1994). https://doi.org/10.1007/978-1-4612-2648-21
8. Rubinstein, A.: Equilibrium in Supergames. Essays in Game Theory. (1994). https://doi.org/10.1007/978-1-4612-2648-22
9. Fudenberg, D., Maskin, E.: The Folk Theorem in Repeated Games with Discounting or with Incomplete Information. Econometrica. **54**, 3, 533–554 (1986). https://doi.org/10.2307/1911307.JSTOR1911307
10. Petrosjan L.A., Pankratova, Y.B.: Construction of Strong Nash Equilibria in a class of infinite nonzero-sum games. Trudy Inst. Mat. Mekh. UrO RAN. **24** (2018)
11. Petrosjan L.A.: Strongly time-consistent differential optimality principles. Vestnik St. Petersburg Univ. Math. **26**, 4, 40–46 (1993)

Part II
Stackelberg Games

Incentive Stackelberg Games for Stochastic Systems

Hiroaki Mukaidani

Abstract Dynamic games with hierarchical structure have been identified as key components of modern control systems that enable the integration of renewable cooperative and/or non-cooperative control such as distributed multi-agent systems. Although the incentive Stackelberg strategy has been admitted as the hierarchical strategy that induces the behavior of the decision maker as that of the follower, the followers optimize their costs under incentives without a specific information. Therefore, leaders succeed in using the required strategy to induce the behavior of their followers. This concept is considered very useful and reliable in some practical cases. In this survey, incentive Stackelberg games for deterministic and stochastic linear systems with external disturbance are addressed. The induced features of the hierarchical strategy in the considered models, including stochastic systems governed by Itô stochastic differential equation, Markov jump linear systems, and linear parameter varying (LPV) systems, are explained in detail. Furthermore, basic concepts based on the H_2/H_∞ control setting for the incentive Stackelberg games are reviewed. Next, it is shown that the required set of strategies can be designed by solving higher-order cross-coupled algebraic Riccati-type equations. Finally, as a partial roadmap for the development of the underdeveloped pieces, some open problems are introduced.

Keywords Incentive Stackelberg games · Nash games · Pareto optimal strategy · H_∞ constraint · Stochastic systems · Markov jump systems · Linear parameter varying systems

H. Mukaidani (✉)
Graduate School of Engineering, Hiroshima University, Higashi-Hiroshima City, Hiroshima, Japan
e-mail: mukaida@hiroshima-u.ac.jp

© Springer Nature Switzerland AG 2020   41
D. Yeung et al. (eds.), *Frontiers in Games and Dynamic Games*, Annals of the
International Society of Dynamic Games 16,
https://doi.org/10.1007/978-3-030-39789-0_3

1 Introduction

Over the past few decades, a considerable amount of research has been conducted on various Stackelberg games. It is well known that Stackelberg game is a hierarchical strategy that involves a first movement by the leader and subsequent movements by followers [1]. The basic concept of the Stackelberg game is that the leader determines a strategy in advance and the followers optimize their costs subject to the leader's strategy. Finally, the leader optimizes his or her own cost based on the strategies of the followers. Recently, useful and reliable results on the Stackelberg games have been obtained by investigating various practical applications. For example, let us consider the scheduling problem in a packet switch operation in a ring architecture (Fig. 1) [2].

Incoming packets on each link are stored in a finite capacity buffer and processed by the central processor when it switches among the links according to a scheduling policy. Packets are discarded when a buffer becomes full. The buffer dynamics of that link are controlled locally based on the information about the state of the buffer when the central processor is serving a particular link. For this real-world situation, a central processor (called leader) has the ability to decide the strategy such that the total throughput is maximized by inducing the local controllers for each link (called follower). Specifically, we define the strategy design problem as choosing a static output feedback control that minimizes the cost of transient around an equilibrium. In this case, the objective function for each player that consists of each term in the integrand penalizes transients on the queue length, queue length rate, and the fluctuation of the loss probability, respectively [3].

The task of designing a strategy for this scheduling problem is challenging. For example, a Stackelberg game model was used to describe the situation in which

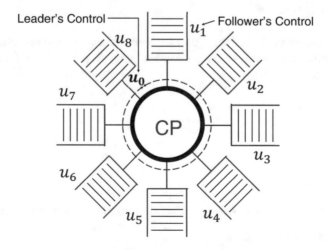

Fig. 1 Scheduling problem in a packet switch operating in a ring architecture

a leader and a follower use the same solar heating system to heat a solar storage and swimming pool, respectively [4]. A Stackelberg game model has been applied to describe electricity trading between a utility company (leader) and multiple electricity users (followers) [5].

In general, Stackelberg strategy is a hierarchical equilibrium solution in Stackelberg games. In [6], an open-loop Stackelberg strategy and a closed-loop feedback Stackelberg strategy were developed. Closed-loop Stackelberg strategies for deterministic systems have been extensively discussed for linear quadratic problems [7]. Recently, in the studies of theory and application of Stackelberg strategy, the growing interest in multi-agent, cooperative, and stochastic systems has led to extensive research in this direction [8–12]. In Stackelberg strategy, the leader's strategy can induce the decision or action of the followers such that the leader's team-optimal solution can be achieved. This kind of strategy is called the incentive Stackelberg strategy, and this has been extensively studied for more than 30 years [13–26]. In [13], incentive problems and corresponding solutions in Stackelberg problems based on the economic literature have been discussed partially. A new approach to obtain the closed-loop Stackelberg solution for a class of two-person nonzero-sum dynamic games characterized by linear quadratic dynamic games was developed in [14]. Another approach to design the nonzero-sum closed-loop Stackelberg strategy under the same conditions such that the leader can achieve the infimum of his/her own criterion was discussed in [15]. In [16], an incentive Stackelberg problem with perfect or partial dynamic information admitting optimal incentive schemes with affine in the available information was addressed. The derivation of causal real-time implementable optimal closed-loop incentive Stackelberg strategies for a general class of discrete and continuous time was discussed in [17]. Sufficient conditions for the incentive Stackelberg strategies in linear quadratic differential games under relaxed conditions on the system parameters were obtained [18]. In [19–21], necessary and sufficient conditions for the existence of a Stackelberg solution for the multi-stage Stackelberg games were given. The linear quadratic incentive Stackelberg game with multi-players in a two-level hierarchy has been studied in [22, 23]. In [24], the three-level incentive Stackelberg strategy in a nonlinear differential game was tackled and the sufficient condition for a linear quadratic differential game was established. In [25], non-cooperative equilibria of many players with multi-levels of hierarchy in decision making was studied. In [26], the team-optimal state-feedback incentive Stackelberg strategy was developed for two-person discrete-time dynamic games that are characterized by linear state dynamics and quadratic cost functions. The unification of these important results in the incentive Stackelberg problems has been provided. However, of all existing problems and corresponding solutions, only a deterministic case has been investigated. Furthermore, the presence of external disturbances and/or unmodeled system dynamics over robust control has not been considered.

Recent advances in deterministic and stochastic systems on robust control theory have resulted in revisiting incentive Stackelberg games [27–38]. In [27], the incentive Stackelberg game for stochastic systems governed by the Itô differential equation in a two-level hierarchy was explained and discussed for the first time.

In [28], the incentive Stackelberg game for deterministic discrete-time systems with deterministic external disturbance was studied. In [29], the study focused on the single-leader-follower incentive Stackelberg game for discrete-time stochastic systems with external deterministic disturbance as an extension of the previous results. In [30, 33, 37], incentive Stackelberg games with multiple leaders and followers were investigated. In particular, a novel concept of incentive possibility has been discussed for a special case [33]. In [31, 35], the incentive Stackelberg games with one leader and multiple followers under the H_∞ constraint for the discrete- and continuous-time stochastic systems have been investigated. In [32], the incentive Stackelberg strategy for the infinite-horizon continuous-time Markov jump stochastic systems with deterministic disturbance was developed. In [34], the incentive Stackelberg game for discrete-time Markov jump stochastic systems with external disturbance by means of static output feedback (SOF) strategy has been addressed. In addition to the above stochastic systems, the H_∞ constraint incentive Stackelberg-Nash and Stackelberg-Pareto strategies for stochastic LPV systems were studied [36, 38].

In this survey, incentive Stackelberg games for a class of stochastic systems with external disturbance are explained. Although several references consider deterministic problems, the stochastic cases are mainly reviewed here. First, after showing a general method for solving the two-level hierarchical games by a simple mathematical example to review this important concept, incentive Stackelberg games with one leader and multiple followers are investigated for a class of continuous-time stochastic linear systems with H_∞ constraint. Unlike the existing ordinary Stackelberg games, the leader is required to design an incentive Stackelberg strategy set that can lead to the leader's team-optimal solution and the follower's Nash equilibrium, and attenuate the external disturbance in the system simultaneously. Second, an infinite-horizon incentive Stackelberg games for a class of Markovian jump linear stochastic systems (MJLSSs) governed by the Itô differential equation are investigated. These games had multiple leaders and followers in a two-level hierarchy and were investigated in line with the generalization of the existing result found in [2]. As another important contribution, a novel concept of incentive possibility is introduced. In contrast to the existing studies, the SOF incentive Stackelberg strategy with H_∞ constraint is studied for the first time. Conversely, a robust incentive Stackelberg game for a class of stochastic LPV systems with multiple decision makers is investigated. The aim in this section is to design a robust incentive Stackelberg strategy in a linear state-feedback format with no scheduling parameters due to the limitation of the real-time control applications. To determine the strategy set and the related incentive of the decision makers, it should be noted that a strategy is designed with a fixed gain and an incentive. Finally, conclusions and suggestions for future work are presented.

Notation The notations used in this survey are fairly standard. I_n denotes the $n \times n$ identity matrix. **col** denotes a column vector. $\|v\|$ denotes the Euclidean norm. $\mathbb{E}[\cdot]$ denotes the expectation operator. $\mathcal{L}^2_{\mathcal{F}}([0, t_f], \mathbb{R}^\ell)$ $(\mathcal{L}^2_{\mathcal{F}}(\mathbb{R}_+, \mathbb{R}^\ell))$ denotes the space of nonanticipative stochastic processes $y(t) \in \mathbb{R}^\ell$ satisfying $\mathbb{E}[\int_0^{t_f} \|y(t)\|^2 dt]$

$(\mathbb{E}[\int_0^\infty \|y(t)\|^2 dt])$. $\mathbb{M}_{n,m}^s$ represents the space of all $\mathbf{A} = (A(1),\ldots,A(s))$ with $A(k)$ being $n \times m$ matrix, $k \in \mathcal{D}$, where $\mathcal{D} = \{1, 2, \ldots, s\}$. The component of $\mathbf{S} + \mathbf{TU}$ is defined as $\mathbf{S} + \mathbf{TU} = (S(1) + T(1)U(1), \ldots, S(s) + T(s)U(s))$. $\mathcal{L}_{\mathcal{F}}^2(\mathbb{R}_+, \mathbb{R}^k)$ denotes the space consisting of all measurable functions $u(t, \omega) : \mathbb{R}_+ \times \Omega \to \mathbb{R}^k$, which is $\mathcal{F}_t$-measurable for every $t \geq 0$, and $\mathbb{E}[\int_0^\infty \|u(t)\|^2 dt \mid r_t = k] < \infty$, $i \in \mathcal{D}$. Γ_i denotes the set of admissible strategies for player P_i, where a strategy set is said to be admissible if the resultant closed-loop system is asymptotically mean-square stable (AMSS). Finally, for an N-tuple $u = (u_1, \ldots, u_N) \in \Gamma_1 \times \cdots \times \Gamma_N$ and for given sets Γ_i, we write

$$u_{-i}^* := (u_1^*, \ldots, u_{i-1}^*, u_i, u_{i+1}^*, \ldots, u_N^*),$$

$$u_{-0i}^* := (u_{01}(u_1^*), \ldots, u_{0(i-1)}(u_{i-1}^*), u_{0i}(u_i), u_{0(i+1)}(u_{i+1}^*), \ldots, u_{0N}(u_N^*))$$

with $u_i \in \Gamma_i$, where the superscript (*) is used in the optimal case. $m\mathrm{e}\text{-}x$ stands for a value of $m \times 10^{-x}$. χ_A denotes indicator function.

2 The Basic Concept of Incentive Stackelberg Game

In order to summarize the incentive Stackelberg game, let us consider two-level hierarchical games with one leader and multiple non-cooperative followers. The hierarchical structure is depicted in Fig. 2. Among multiple players P_i, $i = 0, 1, \ldots, N$, P_0 is considered as the leader and $P_1, \ldots, P_N$ are considered as the followers, under the specification that each follower acts non-cooperatively.

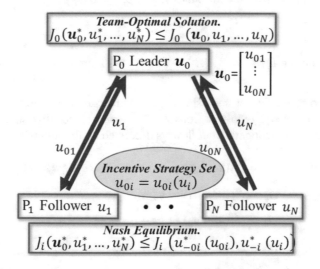

Fig. 2 Incentive Stackelberg hierarchy

The cost functionals of the leader and followers are given by

$$\mathcal{J}_0 = \mathcal{J}_0(\boldsymbol{u}_0, u_1, \ldots, u_N), \tag{1a}$$

$$\mathcal{J}_i = \mathcal{J}_i(\boldsymbol{u}_0, u_1, \ldots, u_N), \ i = 1, \ldots, N, \tag{1b}$$

where $\boldsymbol{u}_0 = \mathbf{col}\begin{bmatrix} u_{01} & \cdots & u_{0N} \end{bmatrix}$.

Definition 1 ([1, 18]) A strategy set $(\boldsymbol{u}_0^*, u_1^*, \ldots, u_N^*)$ is called the team-optimal solution of the game if

$$\mathcal{J}_0(\boldsymbol{u}_0^*, u_1^*, \ldots, u_N^*) \leq \mathcal{J}_0(\boldsymbol{u}_0, u_1, \ldots, u_N) \tag{2}$$

for any $\boldsymbol{u}_0$ and $u_i, i = 1, \ldots, N$.

The framework of the incentive Stackelberg games can be described as follows:

1. The player P_0 announces the following strategy in advance to the players P_i:

$$u_{0i} = u_{0i}(u_i). \tag{3}$$

2. Each player P_i decides his/her own optimal strategy $u_i^*, i = 1, \ldots, N$ under the following Nash equilibrium solution concept, considering the announced strategy of the player P_0.

$$\mathcal{J}_i(\boldsymbol{u}_0^*, u_1^*, \ldots, u_N^*) \leq \mathcal{J}_i(u_{-0i}^*(u_{0i}), u_{-i}^*(u_i)). \tag{4}$$

3. The player P_0 finalizes the incentive Stackelberg strategy

$$u_{0i}^* = u_{0i}^*(u_i^*) \tag{5}$$

for each player P_i, $i = 1, \ldots, N$ so that the team-optimal solution can be achieved.

In order to gain a deeper understanding, a mathematical academic example of a static game with $N = 3$ is solved. Consider the two-level incentive static game with one leader three non-cooperative followers described by the following form:

$$\mathcal{J}_0 = \mathcal{J}_0(\boldsymbol{u}_0, u_1, u_2, u_3) = \boldsymbol{u}^T A_0 \boldsymbol{u} - b_0^T \boldsymbol{u}, \tag{6a}$$

$$\mathcal{J}_1 = \mathcal{J}_1(u_{01}, u_1) = u_1^2 + 2u_{01}^2, \tag{6b}$$

$$\mathcal{J}_2 = \mathcal{J}_2(u_{02}, u_2) = u_2^2 + 3u_{02}^2, \tag{6c}$$

$$\mathcal{J}_3 = \mathcal{J}_3(u_{03}, u_3) = u_3^2 + 4u_{03}^2, \tag{6d}$$

where

$$
A_0 = \begin{bmatrix} 2 & 1 & 1 & 1 & 1 & 1 \\ 1 & 2 & 1 & 1 & 1 & 1 \\ 1 & 1 & 2 & 1 & 1 & 1 \\ 1 & 1 & 1 & 2 & 1 & 1 \\ 1 & 1 & 1 & 1 & 2 & 1 \\ 1 & 1 & 1 & 1 & 1 & 2 \end{bmatrix} > 0, \; b_0 = \begin{bmatrix} 1 \\ 2 \\ -3 \\ 2 \\ 2 \\ -1 \end{bmatrix}, \; u = \begin{bmatrix} u_0 \\ u_1 \\ u_2 \\ u_3 \end{bmatrix}, \; u_0 = \begin{bmatrix} u_{01} \\ u_{02} \\ u_{03} \end{bmatrix}.
$$

The main steps to find the two-level incentive Stackelberg strategies $u_{0i}(u_i)$ of the players P_i $i = 0, 1, \ldots, 3$ are as follows:

1. Using the well-known optimization technique, the team-optimal solution of P_0 can be obtained

$$
u^* = \frac{1}{2} A_0^{-1} b_0 = \begin{bmatrix} u_{01}^* \\ u_{02}^* \\ u_{03}^* \\ u_1^* \\ u_2^* \\ u_3^* \end{bmatrix} = \frac{1}{14} \begin{bmatrix} 4 \\ 11 \\ -24 \\ 11 \\ 11 \\ -10 \end{bmatrix}. \tag{7}
$$

2. Next, let us consider the two-level incentive Stackelberg strategies $u_{0i}(u_i)$ of the form

$$
u_{0i} = u_{0i}(u_i) = u_{0i}^* + \eta_{ii}(u_i - u_i^*), \tag{8}
$$

where η_{ii} $i = 1, \ldots, 3$ are three undetermined coefficients.

Substituting u_i^* and u_{0i}^* from (7) into (8), the following strategy set can be defined.

$$
u_{01} = u_{01}(u_1) = \frac{4}{14} + \eta_{11}\left(u_1 - \frac{11}{14}\right), \tag{9a}
$$

$$
u_{02} = u_{02}(u_2) = \frac{11}{14} + \eta_{22}\left(u_2 - \frac{11}{14}\right), \tag{9b}
$$

$$
u_{03} = u_{03}(u_3) = -\frac{24}{14} + \eta_{33}\left(u_3 + \frac{10}{14}\right). \tag{9c}
$$

3. Solve the following optimization problems:

$$
\min_{u_1} \mathcal{J}_1(u_{01}(u_1), u_{02}(u_2), u_{03}(u_3), u_1, u_2, u_3)
$$

$$
= \min_{u_1}\left[u_1^2 + 2\left\{\frac{4}{14} + \eta_{11}\left(u_1 - \frac{11}{14}\right)\right\}^2\right], \tag{10a}
$$

$$\min_{u_2} \mathcal{J}_2(u_{01}(u_1), u_{02}(u_2), u_{03}(u_3), u_1, u_2, u_3)$$

$$= \min_{u_2} \left[u_2^2 + 3 \left\{ \frac{11}{14} + \eta_{22} \left(u_2 - \frac{11}{14} \right) \right\}^2 \right], \tag{10b}$$

$$\min_{u_3} \mathcal{J}_3(u_{01}(u_1), u_{02}(u_2), u_{03}(u_3), u_1, u_2, u_3)$$

$$= \min_{u_3} \left[u_3^2 + 4 \left\{ -\frac{24}{14} + \eta_{33} \left(u_3 + \frac{10}{14} \right) \right\}^2 \right], \tag{10c}$$

such that (9) holds.

Since $\mathcal{J}_i(u_{01}(u_1), u_{02}(u_2), u_{03}(u_3), u_1, u_2, u_3)$ is strictly convex in u_i, the necessary and sufficient conditions for each problem above are given as follows:

$$\frac{\partial \mathcal{J}_1}{\partial u_1} = 2u_1 + 4\eta_{11} \left\{ \frac{4}{14} + \eta_{11} \left(u_1 - \frac{11}{14} \right) \right\} = 0, \tag{11a}$$

$$\frac{\partial \mathcal{J}_2}{\partial u_2} = 2u_2 + 6\eta_{22} \left\{ \frac{11}{14} + \eta_{22} \left(u_2 - \frac{11}{14} \right) \right\} = 0, \tag{11b}$$

$$\frac{\partial \mathcal{J}_3}{\partial u_3} = 2u_3 + 8\eta_{33} \left\{ -\frac{24}{14} + \eta_{33} \left(u_3 + \frac{10}{14} \right) \right\} = 0. \tag{11c}$$

4. When $u_i = u_i^*, i = 1, \ldots, 3$ are chosen for Eqs. (11), then

$$\eta_{11} = -\frac{7}{4} u_1^* = -\frac{11}{8}, \tag{12a}$$

$$\eta_{22} = -\frac{14}{33} u_2^* = -\frac{1}{3}, \tag{12b}$$

$$\eta_{23} = \frac{7}{48} u_3^* = -\frac{5}{48}. \tag{12c}$$

Therefore, the two-level incentive Stackelberg strategies of P_i are obtained as follows:

$$u_{01} = -\frac{11}{8} u_1 + \frac{153}{112}, \tag{13a}$$

$$u_{02} = -\frac{1}{3} u_2 + \frac{22}{21}, \tag{13b}$$

$$u_{03} = -\frac{5}{48} u_3 - \frac{601}{336}. \tag{13c}$$

In fact, by the strategies (13) announced ahead of time, the cost functionals of P_i will become

$$\mathcal{J}_1(u_0, u_1, u_2, u_3) = u_1^2 + 2\left(-\frac{11}{8}u_1 + \frac{153}{112}\right)^2, \tag{14a}$$

$$\mathcal{J}_2(u_0, u_1, u_2, u_3) = u_2^2 + 3\left(-\frac{1}{3}u_2 + \frac{22}{21}\right)^2, \tag{14b}$$

$$\mathcal{J}_3(u_0, u_1, u_2, u_3) = u_3^2 + 4\left(-\frac{5}{48}u_3 - \frac{601}{336}\right)^2. \tag{14c}$$

By using the incentive (13), the player P_i's optimal decision must be u_i^*. Thus, the strategies $u_{0i}(u_i)$ given by (8) can be attained by the team-optimal condition, they are then indeed the two-level incentive Stackelberg strategies announced by P_0 in this game.

The above-mentioned problem is very difficult to solve if the matrix contains stochastic noise as the disturbance. In such cases, the expected value based on the appropriate stochastic process should be considered.

It should be noted that some follower may not take the optimal strategy subject to the leader's incentive because it would be possible to take the better strategy in the real-life example. Therefore, it is assumed that the follower cannot take their optimal strategies by deviating the leader's incentive such that the leader's team-optimal solution is achieved.

3 Incentive Stackelberg Game with One Leader and Multiple Followers

In this section, the incentive Stackelberg game with one leader and multiple followers for a class of linear stochastic systems governed by the Itô differential equation with external deterministic disturbances is explained by using the previous result in [35]. Since the deterministic disturbances in the systems are also considered in the games, the incentive Stackelberg strategy with the H_∞ constraint is derived. Different from the deterministic incentive Stackelberg games [13, 17, 18, 23, 24, 26], the stochastic incentive Stackelberg game with multiple followers in the systems involving the state-dependent noise and the external deterministic disturbances is discussed.

The structure of the game is depicted in Fig. 3. The game is conducted under the stipulation that the followers act non-cooperatively.

The conditions for the existence of the leader's incentive Stackelberg strategy with the H_∞ constraint are derived on the basis of the existing results for the finite horizon stochastic H_2/H_∞ control problem [39]. It is shown that such a strategy can be obtained by solving a set of the cross-coupled stochastic Riccati differential equations (CCSRDEs); moreover, the strategies for the followers are derived in a way that these strategies satisfy the requirement of the leader's team-optimal solution. It should be noted that the incentive is included in the hierarchical

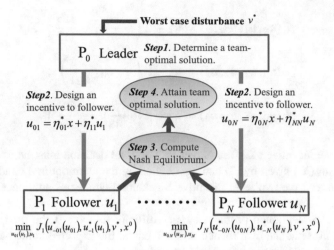

Fig. 3 The incentive Stackelberg strategy with H_∞ constraint

structure, and the followers act to achieve the Nash equilibrium. The stochastic maximum principle plays an important role in the derivation of the incentive Stackelberg strategy. Furthermore, we extend the results to the infinite-horizon case, where, in contrast with the finite-horizon case, the incentive Stackelberg strategy set can be obtained by solving the cross-coupled stochastic algebraic Riccati equations (CCSAREs). In this case, a simple computational algorithm based on the Lyapunov iteration is derived to solve the obtained CCSAREs. Finally, simple numerical examples are provided to illustrate the feasibility and effectiveness of finding the incentive Stackelberg strategy set.

3.1 Definitions and Preliminaries

In this section, we will introduce some definitions and preliminary results on the stochastic H_2/H_∞ control and the LQ control. Firstly, we introduce the team-optimal solution concept [1, 18], which is the essential concept.

Definition 2 ([1, 18]) Let $J_0(u_0, u_1, \ldots, u_N)$ be a given cost function of the leader, where u_0 denotes the leader's control, and u_i, $i = 1, \ldots, N$, denotes the ith follower's control. A control set $(u_0^*, u_1^*, \ldots, u_N^*)$ is called a team-optimal solution if

$$J_0(u_0^*, u_1^*, \ldots, u_N^*) \leq J_0(u_0, u_1, \ldots, u_N) \tag{15}$$

for any u_0 and u_i, $i = 1, 2, \ldots, N$.

It should be noted that if J_0 is quadratic form and strict convex, then a unique optimal solution exists [1].

Secondly, a finite horizon stochastic H_2/H_∞ control problem is introduced [39]. Consider the following linear stochastic system:

$$dx(t) = \big[A(t)x(t) + B(t)u(t) + E(t)v(t)\big]dt + A_p(t)x(t)dw(t),$$

$$x(0) = x^0, \tag{16a}$$

$$z(t) = \mathbf{col}\big[\,C(t)x(t)\ D(t)u(t)\,\big], \quad D^T(t)D(t) = I_m, \tag{16b}$$

where $x(t) \in \mathbb{R}^n$ denotes the state vector. $u(t) \in \mathcal{L}_{\mathcal{F}}^2([0, t_f], \mathbb{R}^m)$ denotes the control input. $v(t) \in \mathcal{L}_{\mathcal{F}}^2([0, t_f], \mathbb{R}^{n_v})$ represents the external deterministic disturbance. $w(t) \in \mathbb{R}$ denotes a one-dimensional standard Wiener process defined in the filtered probability space [6]. $z(t) \in \mathbb{R}^{n_z}$ denotes the controlled output. $A(t)$ and $A_p(t)$ are $n \times n$ dimensional matrices, $B(t)$ is $n \times m$ dimensional matrix, $E(t)$ is $n \times n_v$ dimensional matrix, $C(t)$ is $n_z \times n$ dimensional matrix, and $D(t)$ is $n_z \times m$ dimensional matrix, respectively. Furthermore, all coefficient matrices are time-varying with piecewise-continuous elements.

The finite horizon stochastic H_2/H_∞ control problem can be stated as follows [39]:

Given the disturbance attenuation level $\gamma > 0$, the finite horizon stochastic H_2/H_∞ control problem is to find an optimal state feedback control $u^*(t) \in \mathcal{L}_{\mathcal{F}}^2([0, t_f], \mathbb{R}^m)$ and a worst-case disturbance $v^*(t) \in \mathcal{L}_{\mathcal{F}}^2([0, t_f], \mathbb{R}^{n_v})$ such that

1. when the optimal state feedback control $u^*(t) = K^*(t)x(t)$ is applied,

$$\|\mathcal{L}\|_{[0,t_f]} = \sup_{\substack{v \in \mathcal{L}_{\mathcal{F}}^2([0,\,t_f],\,\mathbb{R}^{n_v}) \\ v \neq 0,\, x^0 = 0}} \frac{\|z\|_{[0,t_f]}}{\|v\|_{[0,t_f]}}$$

$$= \sup_{\substack{v \in \mathcal{L}_{\mathcal{F}}^2([0,\,t_f],\,\mathbb{R}^{n_v}) \\ v \neq 0,\, x^0 = 0}} \frac{\sqrt{J_u(u^*)}}{\sqrt{\mathbb{E}\left[\displaystyle\int_0^{t_f} \|v(t)\|^2 dt\right]}} < \gamma, \tag{17}$$

where $K^*(t)$ is the optimal feedback gain and

$$J_u(u) := \mathbb{E}\left[\int_0^{t_f} \|z(t)\|^2 dt\right]$$

$$= \mathbb{E}\left[\int_0^{t_f} \{x^T(t)C^T(t)C(t)x(t) + u^T(t)u(t)\}dt\right]. \tag{18}$$

2. when the worst-case disturbance $v^*(t) = F^*(t)x(t)$ is applied, $u^*(t)$ minimizes the cost functional $J_u(u, v^*) = J_u(u)$ in (18). It should be noted that $F^*(t)$

denotes the gain of the worst-case disturbance. Here, the so-called worst-case disturbance means that

$$v^*(t) = \arg\min_v J_v(u^*, v), \tag{19}$$

where

$$J_v(u^*, v) = \mathbb{E}\left[\int_0^{t_f} \{\gamma^2 \|v(t)\|^2 - \|z(t)\|^2\} dt\right].$$

It should be noted that if the previous strategy pair (u^*, v^*) exists, then it is said that the finite horizon control has Nash equilibrium solution (u^*, v^*).

Lemma 1 ([39]) *Finite horizon H_2/H_∞ control has solution $(u^*(t), v^*(t)) = (K^*(t)x(t), F^*(t)x(t))$ if and only if the following CCSRDEs (20) have solutions $X(t) \geq 0$ and $Y(t) \geq 0$ on $[0, t_f]$.*

$$-\dot{X}(t) = X(t)A_f(t) + A_f^T(t)X(t) + A_p^T(t)X(t)A_p(t)$$
$$+ X(t)S(t)X(t) + Q_0(t), \tag{20a}$$

$$-\dot{Y}(t) = Y(t)A_f(t) + A_f^T(t)Y(t) + A_p^T(t)Y(t)A_p(t)$$
$$- \gamma^{-2}Y(t)T(t)Y(t) + X(t)S(t)X(t) + Q_0(t), \tag{20b}$$

where

$$X(t_f) = Y(t_f) = 0,$$
$$A_f(t) := A(t) - S(t)X(t) + \gamma^{-2}T(t)Y(t),$$
$$S(t) := B(t)B^T(t),$$
$$T(t) := E(t)E^T(t),$$
$$Q_0(t) := C^T(t)C(t).$$

In this case, the optimal state feedback control and the worst-case disturbance are as given below.

$$u^*(t) = K^*(t)x(t) = -B^T(t)X(t)x(t), \tag{21a}$$

$$v^*(t) = F^*(t)x(t) = \gamma^{-2}E^T(t)Y(t)x(t). \tag{21b}$$

It should be noted that when we consider the infinite-horizon case, the CCSRDEs (20) changes to the CCSAREs. Similar results can be referred in [39] and the related proofs can be found there. The results will be considered when we discuss the infinite-horizon case.

Thirdly, the existing results of the LQ control problem for the stochastic systems in [39] is extended. Let us consider the following finite-time stochastic LQ control problem:

$$\min_u J(u),\tag{22}$$

where

$$J(u) := \frac{1}{2}\mathbb{E}\left[\int_0^{t_f} \{x^T(t)Q(t)x(t) + u^T(t)R(t)u(t) + 2u^T(t)U(t)x(t)\}dt\right],$$

$$Q(t) = Q^T(t) \geq 0, \ R(t) = R^T(t) > 0$$

such that

$$dx(t) = \left[A(t)x(t) + B(t)u(t)\right]dt + A_p(t)x(t)dw(t), \ x(0) = x^0.$$

Lemma 2 ([35]) *Assume that* $Q(t) - U^T(t)R^{-1}(t)U(t) \geq 0$. *Suppose that the following stochastic differential Riccati equation (SDRE) has a solution* $X(t) \geq 0$.

$$-\dot{X}(t) = X(t)\left[A(t) - B(t)R^{-1}(t)U(t)\right]$$

$$+\left[A(t) - B(t)R^{-1}(t)U(t)\right]^T X(t) + A_p^T(t)X(t)A_p(t)$$

$$-X(t)B(t)R^{-1}(t)B^T(t)X(t) + Q(t) - U^T(t)R^{-1}(t)U(t) = 0.\tag{23}$$

Then, the optimal feedback control is given by

$$u(t) = -R^{-1}(t)\left(B^T(t)X(t) + U(t)\right)x(t).\tag{24}$$

It should be noted that the solution is defined in the interval $[0, t_f]$.

Finally, in order to derive our results for the infinite-horizon case, the following facts will be used.

Definition 3 ([35]) Consider the following stochastic system with multiple decision makers:

$$dx(t) = \left[Ax(t) + \sum_{j=0}^N B_j u_j(t)\right]dt + A_p x(t)dw(t).\tag{25}$$

If there exist feedback control $u_i(t) = K_i x(t)$, $i = 0, 1, \ldots, N$ such that for any $x(0) = x^0$, the closed-loop stochastic system (25) is AMSS, the stochastic system with multiple decision makers is called stabilizable.

Definition 4 ([39, 40]) Consider the following stochastic system with measurement equation:

$$dx(t) = Ax(t)dt + A_p x(t)dw(t), \quad x(0) = x^0, \tag{26a}$$

$$z(t) = Cx(t). \tag{26b}$$

If $z(t) \equiv 0, \forall t \geq 0$ implies $x^0 = 0$, $(A, A_p \mid C)$ is called exactly observable.

It may be noted that exact observability of stochastic linear controlled systems has been well documented [40].

In the next two sections, we will formulate the incentive Stackelberg game with H_∞ constraint for the finite-horizon case, and then solve the game formulated.

3.2 Problem Formulation

Consider a linear stochastic system governed by the Itô differential equation defined by

$$dx(t) = \left[A(t)x(t) + \sum_{j=1}^{N} \left[B_{0j}(t)u_{0j}(t) + B_j(t)u_j(t) \right] \right.$$

$$\left. + E(t)v(t) \right] dt + A_p(t)x(t)dw(t), \quad x(0) = x^0, \tag{27a}$$

$$z(t) = \mathbf{col}\left[C(t)x(t) \; u_0(t) \; u_1(t) \cdots u_N(t) \right], \tag{27b}$$

$$u_0(t) = \mathbf{col}\left[u_{01}(t) \cdots u_{0N}(t) \right], \tag{27c}$$

where $B_i(t)$ is $n \times m_i$ dimensional matrix and $B_{0i}(t)$ is $n \times m_{0i}$ dimensional matrix, respectively. Furthermore, these matrices are time-varying with piecewise-continuous elements.

$u_0(t) \in \mathcal{L}_{\mathcal{F}}^2([0, t_f], \mathbb{R}^{m_0})$, $m_0 = \sum_{j=1}^{N} m_{0j}$ with $u_{0j}(t) \in \mathcal{L}_{\mathcal{F}}^2([0, t_f], \mathbb{R}^{m_{0j}})$ denotes the leader's control input. $u_i(t) \in \mathcal{L}_{\mathcal{F}}^2([0, t_f], \mathbb{R}^{m_i}), i = 1, \ldots, N$ denotes the ith follower's control input. In the following, we use P_0 to represent the leader and $P_i, i = 1, \ldots, N$ to represent the ith follower. The definitions of the other variables are the same as those in stochastic system (16).

On the other hand, the cost functions of P_0 and $P_i, i = 1, 2, \ldots, N$ are given by

$$J_0(u_{01}, \ldots, u_{0N}, u_1, \ldots, u_N, v)$$

$$= \frac{1}{2}\mathbb{E}\left[\int_0^{t_f} \left\{ x^T(t)Q_0(t)x(t) + \sum_{i=1}^{N} \left[u_{0i}^T(t)R_{00i}(t)u_{0i}(t) \right.\right.\right.$$

$$\left.\left.\left. + u_i^T(t)R_{0i}(t)u_i(t) \right] \right\} dt \right], \tag{28a}$$

$$J_i(u_{01}, \ldots, u_{0N}, u_1, \ldots, u_N, v)$$

$$= \frac{1}{2}\mathbb{E}\left[\int_0^{t_f} \{x^T(t)Q_i(t)x(t) + u_{0i}^T(t)R_{0ii}(t)u_{0i}(t)\right.$$

$$\left. + u_i^T(t)R_{ii}(t)u_i(t)\}dt\right], \quad i = 1, \ldots, N, \tag{28b}$$

where $Q_0(t) := C^T(t)C(t)$, $Q_i(t) = Q_i^T(t) \geq 0$, $R_{00i}(t) = R_{00i}^T(t) > 0$, $R_{0i}(t) = R_{0i}^T(t) \geq 0$, $R_{0ii}(t) = R_{0ii}^T(t) \geq 0$, and $R_{ii}(t) = R_{ii}^T(t) > 0$, $i = 1, \ldots, N$ are piece-wise continuous functions of time on the fixed interval $[0, t_f]$.

The leader's strategy to which optimal response of the follower (obtained by minimization of his own cost function) coincides with team-optimal strategy is called incentive Stackelberg strategy, thus yielding the optimal team-trajectory [18]. In fact, the game process to determine the incentive Stackelberg strategy set is as follows [14, 18]:

1. The team-optimal strategy is computed

$$u_{0i}^*(t) = K_{0i}^*(t)x(t), \tag{29a}$$

$$u_i^*(t) = K_i^*(t)x(t), \quad i = 1, \ldots, N, \tag{29b}$$

   where $K_{0i}^*(t)$ and $K_i^*(t)$ denote the gains of the team-optimal strategy.
2. The leader announces a strategy ahead of time to the followers with the following feedback pattern:

$$u_{0i}(t) = u_{0i}(t, x(t), u_i) = u_{0i}^*(t) + \eta_{ii}(t)\big(u_i(t) - u_i^*(t)\big)$$

$$= K_{0i}^*(t)x(t) + \eta_{ii}(t)\big(u_i(t) - K_i^*(t)x(t)\big)$$

$$= \eta_{0i}(t)x(t) + \eta_{ii}(t)u_i(t), \quad i = 1, \ldots, N, \tag{30}$$

   where $\eta_{0i}(t) = K_{0i}^*(t) - \eta_{ii}(t)K_i^*(t) \in \mathbb{R}^{m_{0i} \times n}$ and $\eta_{ii}(t) \in \mathbb{R}^{m_{0i} \times m_i}$ are strategy parameter matrices. Moreover, their components are piece-wise continuous functions of time on the interval $[0, t_f]$.
3. The followers determine their strategies to achieve a Nash equilibrium by responding to the announced strategy of the leader.
4. The leader determines the incentive Stackelberg strategy

$$u_{0i}^*(t) = u_{0i}^*(t, x(t), u_i^*) = \eta_{0i}^*(t)x(t) + \eta_{ii}^*(t)u_i^*(t) \tag{31}$$

for $i = 1, \ldots, N$ to achieve the team-optimal solution $(u_0^*, u_1^*, \ldots, u_N^*)$, which is associated with the Nash equilibrium strategy $u_i^*(t)$ for $i = 1, \ldots, N$ of the followers.

The incentive Stackelberg game under the H_∞ constraint with one leader and multiple followers is formulated as follows:

For any given $\gamma > 0$, $0 < t_f < \infty$, find an incentive strategy (31) and a state feedback control $u_i^*(t) = K_i^*(t)x(t) \in \mathcal{L}_{\mathcal{F}}^2([0, t_f], \mathbb{R}^{m_i})$, $i = 1, \ldots, N$ that are associated with a worst-case disturbance $v^*(t) = F_\gamma^*(t)x(t) \in \mathcal{L}_{\mathcal{F}}^2([0, t_f], \mathbb{R}^{m_v})$ such that

1. the stochastic system (27) attains the team-optimal condition (32a) with H_∞ constraint condition (32b)

$$J_0(u_{01}^*, \ldots, u_{0N}^*, u_1^*, \ldots, u_N^*, v^*)$$

$$= \min_{u_{01}, \ldots, u_{0N}, u_1, \ldots, u_N} J_0(u_{01}, \ldots, u_{0N}, u_1, \ldots, u_N, v^*), \qquad (32a)$$

$$0 \leq J_v(u_{01}^*, \ldots, u_{0N}^*, u_1^*, \ldots, u_N^*, v^*)$$

$$\leq J_v(u_{01}^*, \ldots, u_{0N}^*, u_1^*, \ldots, u_N^*, v), \qquad (32b)$$

where

$$J_v(u_{01}, \ldots, u_{0N}, u_1, \ldots, u_N, v) = \mathbb{E}\left[\int_0^{t_f} \{\gamma^2 \|v(t)\|^2 - \|z(t)\|^2\}dt\right],$$

$$\|z(t)\|^2 = x^T(t)Q_0(t)x(t) + \sum_{j=0}^N u_j^T(t)u_j(t)$$

for $\forall v(t) \neq 0 \in \mathcal{L}_{\mathcal{F}}^2([0, t_f], \mathbb{R}^{m_v})$,

2. the control set $(u_{0i}^*, u_i^*) \in \mathbb{R}^{m_{0i}+m_i}$, $i = 1, \ldots, N$ satisfies the inequality:

$$J_i^* = J_i(u_{01}^*(u_1^*), \ldots, u_{0N}^*(u_N^*), u_1^*, \ldots, u_N^*, v^*)$$

$$\leq J_i(u_{-0i}^*(u_{0i}), u_{-i}^*(u_i), v^*)$$

$$= J_i((u_{01}^*(u_1^*), \ldots, u_{0(i-1)}^*(u_{i-1}^*), u_{0i}(u_i), u_{0(i+1)}^*(u_{i+1}^*), \ldots, u_{0N}^*(u_N^*),$$

$$(u_1^*, \ldots, u_{i-1}^*, u_i, u_{i+1}^*, \ldots, u_N^*), v^*). \qquad (33)$$

It should be noted that $F_\gamma^*(t)$ denotes the gains of the worst-case disturbance. Furthermore, it should also be noted that the leader's strategy u_{0i} depends on the follower's strategy u_i because the incentive strategy set (30) holds.

It is obvious that inequality (33) defines a Nash equilibrium [1, 41], that is, the strategy set $(u_{0i}^*, u_i^*) \in \mathbb{R}^{m_{0i}+m_i}$, $i = 1, \ldots, N$ constitute both the team-optimal incentive Stackelberg strategy set with H_∞ constraint of the leader and the Nash equilibrium strategies of the followers.

3.3 Main Results

Let us define the space of admissible strategy Γ_i for P_i, $i = 0, 1, \ldots, N$. For each pair $(u_{0i}, u_i) \in \Gamma_0 \times \Gamma_i$, it is supposed that the linear stochastic systems (27) has a unique solution on $[0, t_f]$ for all initial state x^0 and the values of J_i are well defined. Firstly, the team-optimal solution set with the H_∞ constraint (u_{0i}^*, u_i^*, v^*) is derived. By centralizing the control inputs in the stochastic system (27), the following centralized stochastic systems can be obtained:

$$dx(t) = \left[A(t)x(t) + B_c(t)u_c(t) + E(t)v(t) \right]dt$$

$$+ A_p(t)x(t)dw(t), \quad x(0) = x^0, \tag{34a}$$

$$z(t) = \mathbf{col}\left[C(t)x(t) \; u_c(t) \right], \tag{34b}$$

where

$$B_c(t) := \left[B_0(t) \; B_1(t) \; \cdots \; B_N(t) \right],$$

$$B_0(t) := \left[B_{01}(t) \; \cdots \; B_{0N}(t) \right],$$

$$u_c(t) := \mathbf{col}\left[u_0(t) \; u_1(t) \; \cdots \; u_N(t) \right].$$

Furthermore, the cost functional (28a) can be changed as:

$$J_0(u_{01}, \ldots, u_{0N}, u_1, \ldots, u_N, v)$$

$$= \frac{1}{2}\mathbb{E}\left[\int_0^{t_f} \left\{ x^T(t)Q_0(t)x(t) + u_c^T(t)R_c(t)u_c(t) \right\}dt \right], \tag{35}$$

where

$$Q_0(t) := C^T(t)C(t),$$

$$R_c(t) := \mathbf{block\ diag}\left(R_{00}(t) \; R_{01}(t) \; \cdots \; R_{0N}(t) \right),$$

$$R_{00}(t) := \mathbf{block\ diag}\left(R_{001}(t) \; \cdots \; R_{00N}(t) \right).$$

By using Lemma 1, the following conditions via the CCSRDEs can be obtained:

$$-\dot{P}(t) = P(t)A_c(t) + A_c^T(t)P(t) + A_p^T(t)P(t)A_p(t)$$

$$+ P(t)S_c(t)P(t) + Q_0(t), \tag{36a}$$

$$-\dot{W}(t) = W(t)A_c(t) + A_c^T(t)W(t) + A_p^T(t)W(t)A_p(t)$$

$$- \gamma^{-2}W(t)T(t)W(t) + P(t)S_c(t)P(t) + Q_0(t), \tag{36b}$$

where

$$P(t_f) = W(t_f) = 0,$$

$$A_c := A(t) - S_c(t)P(t) + \gamma^{-2}T(t)W(t),$$

$$S_c(t) := B_c(t)R_c^{-1}(t)B_c^T(t).$$

Furthermore, if $P(t) \geq 0$ and $W(t) \geq 0$ for $\forall t \in [0, t_f]$ hold, the strategy set is as given below.

$$u_c^*(t) = \mathbf{col}\left[u_0^*(t)\, u_1^*(t) \cdots u_N^*(t) \right]$$

$$= K_c^*(t)x(t) = -R_c^{-1}(t)B_c^T(t)P(t)x(t), \tag{37a}$$

$$v^*(t) = F_\gamma^*(t)x(t) = \gamma^{-2}E^T(t)W(t)x(t), \tag{37b}$$

where $K_c^*(t)$ is the optimal feedback gain of the team-optimal strategy and

$$u_0^* := \mathbf{col}\left[u_{01}^*(t) \cdots u_{0N}^*(t) \right],$$

$$u_{0i}^*(t) = K_{0i}^*(t)x(t) = -R_{00i}^{-1}(t)B_{0i}^T(t)P(t)x(t),$$

$$u_i^*(t) = K_i^*(t)x(t) = -R_{0i}^{-1}(t)B_i^T(t)P(t)x(t).$$

Secondly, the non-cooperative Nash strategy set for the followers is derived. Define the following matrix functions:

$$\eta_{0i}(t) = K_{0i}^*(t) - \eta_{ii}(t)K_i^*(t)$$

$$= -R_{00i}^{-1}(t)B_{0i}^T(t)P(t) + \eta_{ii}(t)R_{0i}^{-1}(t)B_i^T(t)P(t). \tag{38}$$

For given $\eta_{ii}(t)$, $i = 1, \ldots, N$ which satisfy (31), let us consider the following optimization problem:

$$\min_{u_{0i}(u_i),u_i} J_i\left(u_{-0i}^*(u_{0i}),\, u_{-i}^*(u_i),\, v^*\right)$$

$$= \min_{u_i} J_i\left(u_{-0i}^*\left(\eta_{0i}(t)x(t) + \eta_{ii}(t)u_i(t)\right),\, u_{-i}^*(u_i),\, v^*\right), \tag{39}$$

such that

$$dx(t) = F(t, x, u_i, \eta_{ii})dt + A_p(t)x(t)dw(t), \quad x(0) = x^0, \tag{40}$$

where

$$F(t, x, u_i, \eta_{ii})$$

$$:= A(t)x(t) + B_{0i}(t)\big[\eta_{0i}(t)x(t) + \eta_{ii}(t)u_i(t)\big] + B_i(t)u_i(t)$$

$$+ \sum_{j=1, j\neq i}^{N} \Big[B_{0j}(t)\big[\eta_{0j}(t)x(t) + \eta_{jj}(t)u_j^*(t)\big] + B_j(t)u_j^*(t) \Big]$$

$$+\gamma^{-2}T(t)W(t)x(t).$$

In order to apply the maximum principle to this optimization problem, the following Hamiltonian is defined:

$$H_i = \frac{1}{2}\Big[x^T(t)Q_i(t)x(t) + \big[\eta_{0i}(t)x(t) + \eta_{ii}(t)u_i(t)\big]^T R_{0ii}(t)$$

$$\times \big[\eta_{0i}(t)x(t) + \eta_{ii}(t)u_i(t)\big] + u_i^T(t)R_{ii}(t)u_i(t) \Big]$$

$$+\alpha_i^T(t)F(t, x, u_i, \eta_{ii}) + \beta_i^T(t)A_p(t)x(t), \quad i = 1, \ldots, N. \tag{41}$$

Hence, we have the following equations:

$$d\alpha_i(t) = -\frac{\partial H_i}{\partial x}dt + \beta_i(t)dw(t)$$

$$= -\Big[\Big(Q_i(t) + \eta_{0i}^T(t)R_{0ii}(t)\eta_{0i}(t) \Big)x(t) + \eta_{0i}^T(t)R_{0ii}(t)\eta_{ii}(t)u_i(t)$$

$$+\big[A(t) + B_{0i}(t)\eta_{0i}(t)\big]^T \alpha_i(t)$$

$$+ \sum_{j=1, j\neq i}^{N} \Big[B_{0j}(t)\eta_{0j}(t) + \bar{B}_j(t)\Big(\frac{\partial u_j^*(t)}{\partial x}\Big) \Big]^T \alpha_i(t)$$

$$+\gamma^{-2}W(t)T(t)\alpha_i(t) + A_p^T(t)\beta_i(t) \Big]dt + \beta_i(t)dw(t), \tag{42a}$$

$$\frac{\partial H_i}{\partial u_i(t)} = \eta_{ii}^T(t)R_{0ii}(t)[\eta_{0i}(t)x(t) + \eta_{ii}(t)u_i(t)]$$

$$+R_{ii}(t)u_i(t) + \bar{B}_i^T(t)\alpha_i(t) = 0, \tag{42b}$$

where $\bar{B}_i(t) := B_i(t) + B_{0i}(t)\eta_{ii}(t)$.

It should be ensured that $u_{0i}^*(t)$, $u_i^*(t)$, and $v^*(t)$ satisfy the H_∞ constraint team-optimal solution of (37). Without loss of generality, since we consider the linear stochastic system, it is assumed that

$$\alpha_i(t) = Z_i(t)x(t). \tag{43}$$

Then, from (42b), we have

$$u_i(t) = u_i^\dagger(t) = K_i^\dagger(t)x(t)$$

$$= -\hat{R}_{ii}^{-1}(t)\left[\eta_{ii}^T(t)R_{0ii}(t)\eta_{0i}(t) + \bar{B}_i^T(t)Z_i(t)\right]x(t), \qquad (44)$$

where $K_i^\dagger(t)$ denotes the optimal feedback gain under the incentive and

$$\hat{R}_{ii}(t) := R_{ii}(t) + \eta_{ii}^T(t)R_{0ii}(t)\eta_{ii}(t).$$

It should be noted that $\hat{R}_{ii}(t)$ is nonsingular because of $R_{ii}(t) > 0$. Then, we assume that the following condition holds:

$$u_i^*(t) = K_i^*(t)x(t) = u_i^\dagger(t) = K_i^\dagger(t)x(t). \qquad (45)$$

Namely, we have

$$K_i^*(t) = -R_{0i}^{-1}(t)B_i^T(t)P(t) = K_i^\dagger(t)$$

$$= -\hat{R}_{ii}^{-1}(t)\left[\eta_{ii}^T(t)R_{0ii}(t)\eta_{0i}(t) + \bar{B}_i^T(t)Z_i(t)\right]. \qquad (46)$$

Therefore, from Eq. (46) we have

$$\eta_{ii}^T(t)\left(B_{0i}^T(t)Z_i(t) - R_{0ii}(t)R_{00i}^{-1}(t)B_{0i}^T(t)P(t)\right)$$

$$= R_{ii}(t)R_{0i}^{-1}(t)B_i^T(t)P(t) - B_i^T(t)Z_i(t), \quad i = 1,\dots,N. \qquad (47)$$

It should be noted that $\eta_{0i}(t) - \eta_{ii}(t)R_{0i}^{-1}(t)B_i^T(t)P(t) = -R_{00i}^{-1}(t)B_{0i}^T(t)P(t)$ of (38) is used.

Furthermore, stochastic system (42a) can be changed as:

$$d\alpha_i(t) = -\left[\left(Q_i(t) + \eta_{0i}^T(t)R_{0ii}(t)\left[\eta_{0i}(t) + \eta_{ii}(t)K_i^*(t)\right]\right.\right.$$

$$+ \left[A(t) + \sum_{j=1}^N B_{0j}(t)\eta_{0j}(t) + \gamma^{-2}T(t)W(t)\right]^T Z_i(t)$$

$$\left.\left. + \sum_{j=1, j\neq i}^N K_j^{*T}(t)\bar{B}_j^T(t)Z_i(t)\right)x(t) + A_p^T(t)\beta_i(t)\right]dt\beta_i(t)dw(t).$$

$$(48)$$

On the other hand, from (43), we have

$$
d\alpha_i(t) = \frac{\partial \alpha_i(t)}{\partial t} dt + \frac{\partial \alpha_i(t)}{\partial x(t)} dx(t)
$$

$$
= \left[\dot{Z}_i(t) + Z_i(t) \left(A(t) \right. \right.
$$

$$
+ \sum_{j=1}^{N} \left[B_{0j}(t) \left[\eta_{0j}(t) + \eta_{jj}(t) K_j^*(t) \right] + B_j(t) K_j^*(t) \right]
$$

$$
\left. \left. + \gamma^{-2} T(t) W(t) \right) \right] x(t) dt + Z_i(t) A_p(t) x(t) dw(t), \tag{49}
$$

where

$$
d\alpha_{ik}(t) = \frac{\partial \alpha_{ik}}{\partial t} dt + \sum_{p=1}^{n} \frac{\partial \alpha_{ik}}{\partial x_p} dx_p + \frac{1}{2} \sum_{q=1}^{n} \sum_{p=1}^{n} \frac{\partial^2 \alpha_{ik}}{\partial x_q \partial x_p} dx_p dx_q,
$$

$$
x(t) = \begin{bmatrix} x_1(t) \\ \vdots \\ x_n(t) \end{bmatrix}, \quad \alpha_i(t) = \begin{bmatrix} \alpha_{i1}(t) \\ \vdots \\ \alpha_{in}(t) \end{bmatrix}, \quad \frac{\partial^2 \alpha_{ik}}{\partial x_q \partial x_p} = 0.
$$

By the term-wise comparison between (48) and (49), we have

$$
-\dot{Z}_i(t) = Z_i(t) \hat{A}(t) + \hat{A}^T(t) Z_i(t) + A_p^T(t) Z_i(t) A_p(t)
$$

$$
+ Z_i(t) \bar{B}_i(t) \hat{R}_{ii}^{-1}(t) \bar{B}_i^T(t) Z_i(t) + \hat{Q}_i(t)
$$

$$
- \eta_{0i}^T(t) R_{0ii}(t) \eta_{ii}(t) \hat{R}_{ii}^{-1}(t) \eta_{ii}^T(t) R_{0ii}(t) \eta_{0i}(t), \tag{50}
$$

where

$$
\hat{A}(t) := A(t) - \sum_{j=1}^{N} B_{0j}(t) R_{00j}^{-1}(t) B_{0j}^T(t) P(t)
$$

$$
- \sum_{j=1}^{N} B_j(t) R_{0j}^{-1}(t) B_j^T(t) P(t) + \gamma^{-2} T(t) W(t),
$$

$$
\hat{Q}_i(t) := Q_i(t) + \eta_{0i}^T(t) R_{0ii}(t) \eta_{0i}(t).
$$

We are now in a position to state the main result for the incentive Stackelberg game with H_∞ constraint.

Theorem 1 ([35]) *Suppose that the solutions of the CCSRDEs (36), the matrix algebraic equations (MAEs) (47), and the SDREs (50) exist. Then the strategy set (31) associated with $u_i^*(t) = K_i^*(t) x(t)$ for $i = 1, \ldots, N$ constitute the incentive*

Stackelberg strategy set with H_∞ constraint as formulated in Section III, where $\eta_{0i}^*(t)$ and $\eta_{ii}^*(t)$ are determined by using (38) and (47), respectively.

It should be noted that $\eta_{0i}^*(t)$ and $\eta_{ii}^*(t)$ denote the required incentive.

Remark 1 It should be noted that the unknown variables of the SDREs (50) are $Z_i(t)$ and $\eta_{ii}(t)$. However, we have the relation from (47). Hence, it is possible to obtain the solutions $Z_i(t)$ and $\eta_{ii}(t)$ by combining Eqs. (38), (47), and (50).

3.4 Infinite-Horizon Case

In this section, the infinite-horizon incentive Stackelberg game with multiple followers subject to the H_∞ constraint is explained. Consider a time-invariant linear stochastic system governed by the Itô stochastic differential equation:

$$dx(t) = \left[Ax(t) + \sum_{j=1}^{N} \left[B_{0j} u_{0j}(t) + B_j u_j(t) \right] + Ev(t) \right] dt$$

$$+ A_p x(t) dw(t), \quad x(0) = x^0, \tag{51a}$$

$$z(t) = \mathbf{col} \left[Cx(t)\ u_0(t)\ u_1(t)\ \cdots\ u_N(t) \right], \tag{51b}$$

$$u_0(t) = \mathbf{col} \left[u_{01}(t)\ \cdots\ u_{0N}(t) \right], \tag{51c}$$

where the dimension of all coefficient matrices is the same as the stochastic system (27) and they are constant matrices.

Moreover, the cost functions are defined as follows:

$$J_0(u_{01}, \ldots, u_{0N}, u_1, \ldots, u_N, v)$$

$$= \frac{1}{2} \mathbb{E} \left[\int_0^\infty \left\{ x^T(t) Q_0 x(t) + \sum_{i=1}^{N} \left[u_{0i}^T(t) R_{00i} u_{0i}(t) \right. \right. \right.$$

$$\left. \left. \left. + u_i^T(t) R_{0i} u_i(t) \right] \right\} dt \right], \tag{52a}$$

$$J_i(u_{01}, \ldots, u_{0N}, u_1, \ldots, u_N, v)$$

$$= \frac{1}{2} \mathbb{E} \left[\int_0^\infty \left\{ x^T(t) Q_i x(t) + u_{0i}^T(t) R_{0ii} u_{0i}(t) + u_i^T(t) R_{ii} u_i(t) \right\} dt \right], \tag{52b}$$

$$J_v(u_{01}, \ldots, u_{0N}, u_1, \ldots, u_N, v)$$

$$= \mathbb{E} \left[\int_0^\infty \left\{ \gamma^2 \|v(t)\|^2 - \|z(t)\|^2 \right\} dt \right], \tag{52c}$$

where $Q_0 := C^T C$, $Q_i = Q_i^T \geq 0$, $R_{00i} = R_{00i}^T > 0$, $R_{0i} = R_{0i}^T \geq 0$, $R_{0ii} = R_{0ii}^T \geq 0$, and $R_{ii} = R_{ii}^T > 0$, $i = 1, \ldots, N$.

The infinite-horizon incentive Stackelberg game with H_∞ constraint is formulated as follows:

For any given $\gamma > 0$, find the incentive strategy

$$u_{0i}(t) = u_{0i}^*(t) = \eta_{0i}^* x(t) + \eta_{ii}^* u_i(t), \quad i = 1, \ldots, N, \tag{53}$$

where η_{0i}^* and η_{ii}^* are the parameters to be determined, and a state feedback strategy $u_i^*(t) = K_i^* x(t) \in \mathcal{L}_{\mathcal{F}}^2(\mathbb{R}_+, \mathbb{R}^{m_i})$, $i = 1, \ldots, N$ that is associated with a worst-case disturbance $v^*(t) \in \mathcal{L}_{\mathcal{F}}^2(\mathbb{R}_+, \mathbb{R}^{m_v})$ such that

1. the stochastic system (51) attains the team-optimal condition (54a) with H_∞ constraint condition (54b)

$$
\begin{aligned}
J_0(u_{01}^*, &\ldots, u_{0N}^*, u_1^*, \ldots, u_N^*, v^*) \\
&= \min_{u_{01}, \ldots, u_{0N}, u_1, \ldots, u_N} J_0(u_{01}, \ldots, u_{0N}, u_1, \ldots, u_N, v^*), \tag{54a}
\end{aligned}
$$

$$
\begin{aligned}
0 \leq J_v(u_{01}^*, &\ldots, u_{0N}^*, u_1^*, \ldots, u_N^*, v^*) \\
&\leq J_v(u_{01}^*, \ldots, u_{0N}^*, u_1^*, \ldots, u_N^*, v), \tag{54b}
\end{aligned}
$$

2. the control set $(u_{0i}^*, u_i^*) \in \mathbb{R}^{m_{0i}+m_i}$, $i = 1, \ldots, N$ satisfies the following Nash equilibrium condition:

$$
\begin{aligned}
J_i^* &= J_i(u_{01}^*, \ldots, u_{0N}^*, u_1^*, \ldots, u_N^*, v^*) \\
&\leq J_i(u_{-0i}^*(u_{0i}), u_{-i}^*(u_i), v^*). \tag{55}
\end{aligned}
$$

It should be noted that if inequality (54b) holds, then we have

$$\|\mathcal{L}\|_\infty = \sup_{\substack{v \in \mathcal{L}_{\mathcal{F}}^2(\mathbb{R}_+, \mathbb{R}^{n_v}) \\ v \neq 0, \, x^0 = 0}} \frac{\|z\|_2}{\|v\|_2} < \gamma. \tag{56}$$

Firstly, we are interested to find the solution for the infinite-horizon case. For this purpose, we rearrange the system (51) as the same as the finite-horizon case.

$$dx(t) = [Ax(t) + B_c u_c + Ev(t)] dt + A_p x(t) dw(t), \quad x(0) = x^0, \tag{57a}$$

$$z(t) = \mathbf{col}\left[Cx(t)\ u_c(t) \right], \tag{57b}$$

where

$$B_c := \left[B_0\ B_1\ \cdots\ B_N \right],$$

$$B_0 := \begin{bmatrix} B_{01} & \cdots & B_{0N} \end{bmatrix}.$$

The cost functional (52a) is also rearranged as follows:

$$J_{0c}(u_c) = \frac{1}{2}\mathbb{E}\left[\int_0^\infty \left\{x^T(t)Q_0x(t) + u_c^T(t)R_cu_c(t)dt\right\}\right], \tag{58}$$

where

$$R_c := \textbf{block diag}\left(R_{00}\ R_{01}\ \cdots\ R_{0N}\right),$$

$$R_{00} := \textbf{block diag}\left(R_{001}\ \cdots\ R_{00N}\right).$$

Therefore, for the team-optimal solution with the H_∞ constraint, the following result can be obtained from [39]:

For a given $\gamma > 0$, suppose the CCSAREs (59):

$$PA_c + A_c^T P + A_p^T PA_p + PS_cP + Q_0 = 0, \tag{59a}$$

$$WA_c + A_c^T W + A_p^T WA_p - \gamma^{-2}WTW + PS_cP + Q_0 = 0, \tag{59b}$$

with

$$A_c := A - S_cP + \gamma^{-2}TW,$$

$$S_c := B_cR_c^{-1}B_c^T,$$

$$T := EE^T$$

have a pair of solutions $P > 0$ and $W > 0$. If $(A,\ A_p\ |\ C)$ and $(A + \gamma^{-2}TW,\ A_p\ |\ C)$ are exactly observable, then the stochastic team-optimal strategy with H_∞ constraint problem admits a pair of solutions:

$$u_c^*(t) = K_c^*x(t) = -R_c^{-1}B_c^T P^*x(t) = \begin{bmatrix} K_{01}^* \\ \vdots \\ K_{0N}^* \\ K_1^* \\ \vdots \\ K_N^* \end{bmatrix} x(t), \tag{60a}$$

$$v^*(t) = F_\gamma^*x(t) = \gamma^{-2}E^T W^*x(t), \tag{60b}$$

where $P = P^* > 0$ and $W = W^* > 0$ are the solution set of (59).

Secondly, by substituting the leader's incentive Stackelberg strategy

$$u_{0i}(t) = \eta_{0i}x(t) + \eta_{ii}u_i(t) \tag{61}$$

into the stochastic system (51) and the followers' cost functional (52b), the following stochastic algebraic Riccati equations (SAREs) (62a) and MAEs (62b) can be obtained:

$$Z_i\hat{A} + \hat{A}^T Z_i + A_p^T Z_i A_p + Z_i \bar{B}_i \hat{R}_{ii}^{-1} \bar{B}_i^T Z_i$$

$$+ \hat{Q}_i - \eta_{0i}^T R_{0ii} \eta_{ii} \hat{R}_{ii}^{-1} \eta_{ii}^T R_{0ii} \eta_{0i} = 0, \tag{62a}$$

$$\eta_{ii}^T \left(B_{0i}^T Z_i - R_{0ii} R_{00i}^{-1} B_{0i}^T P \right) = R_{ii} R_{0i}^{-1} B_i^T P - B_i^T Z_i, \tag{62b}$$

where $i = 1, \ldots, N$,

$$\hat{A} := A - \sum_{j=1}^{N} B_{0j} R_{00j}^{-1} B_{0j}^T P - \sum_{j=1}^{N} B_j R_{0j}^{-1} B_j^T P + \gamma^{-2} T W,$$

$$\bar{B}_i := B_i + B_{0i}\eta_{ii},$$

$$\hat{Q}_i := Q_i + \eta_{0i}^T R_{0ii}\eta_{0i},$$

$$\eta_{0i} := -R_{00i}^{-1} B_{0i}^T P + \eta_{ii} R_{0i}^{-1} B_i^T P,$$

$$\hat{R}_{ii} := R_{ii} + \eta_{ii}^T R_{0ii}\eta_{ii}.$$

From the above discussions, we can now state another important result.

Theorem 2 ([35]) *Suppose that the stochastic system (51) is stabilizable, the CCSAREs (59) admit the solution set (P^*, W^*) such that $P = P^* > 0$, $W = W^* > 0$, and $(A, A_p \mid C)$ and $(A + \gamma^{-2}T W^*, A_p \mid C)$ are exactly observable. If the SAREs (62a) and the MAEs (62b) have the solution set (Z_i^*, η_{ii}^*) such that $Z_i = Z_i^* > 0$, the following strategy set (63) and (64) is the incentive Stackelberg strategy set of the leader and the followers under the H_∞ constraint.*

$$u_{0i}^*(t) = \eta_{0i}^* x(t) + \eta_{ii}^* u_i(t), \; i = 1, \ldots, N, \tag{63}$$

where $\eta_{0i}^* := -R_{00i}^{-1} B_{0i}^T P^* + \eta_{ii}^* R_{0i}^{-1} B_i^T P^*$.

Moreover, the incentive strategies (63) and the state feedback strategies (64) form the team-optimal solution for J_0.

$$u_i^*(t) = u_i^\dagger(t) = -\hat{R}_{ii}^{-1}\left(\eta_{ii}^{*T} R_{0ii}\eta_{0i}^* + \bar{B}_i^T Z_i^*\right)x(t), \; i = 1, \ldots, N. \tag{64}$$

One may solve the following CCSAREs (65) instead of the independent SAREs (62):

$$\Xi_i \tilde{A} + \tilde{A}^T \Xi_i + A_p^T \Xi_i A_p + \Xi_i \bar{B}_i \hat{R}_{ii}^{-1} \bar{B}_i^T \Xi_i$$

$$+ \hat{Q}_i - \eta_{0i}^T R_{0ii} \eta_{ii} \hat{R}_{ii}^{-1} \eta_{ii}^T R_{0ii} \eta_{0i} = 0, \tag{65a}$$

$$\eta_{ii}^T \left(B_{0i}^T \Xi_i - R_{0ii} R_{00i}^{-1} B_{0i}^T P \right) = R_{ii} R_{0i}^{-1} B_i^T P - B_i^T \Xi_i, \tag{65b}$$

where

$$\tilde{A} := A + \sum_{j=1}^{N} B_{0j} \eta_{0j} + \sum_{j=1}^{N} \bar{B}_j \tilde{K}_j + \gamma^{-2} T W,$$

$$\hat{Q}_i := Q_i + \eta_{0i}^T R_{0i} \eta_{0i},$$

$$\tilde{K}_i := -\hat{R}_{ii}^{-1} \left(\eta_{ii}^T R_{0ii} \eta_{0i} + \bar{B}_i^T \Xi_i \right).$$

It should be noted that $\tilde{A}$ in CCSAREs (65) depends on Ξ_i, $i = 1, \ldots, N$. Hence, these equations are the standard CCSARE of Nash game because Ξ_i, $i = 1, \ldots, N$ exist in $\hat{A}$. There is no doubt that the computed feedback gains K_i^* and $\tilde{K}_i$ from the CCSAREs (62a) and the SAREs (65a) are the same, respectively. This important feature comes from the team-optimal solution. Namely, since matrices Z_i and Ξ_i are constrained by Eqs. (65a) and (65b) for the constant matrix P of (59a), respectively, the same solutions are obtained. In fact, it can be observed that the same solutions are yielded in the numerical example and it will be proved in Appendix. Finally, it should also be noted that SAREs (62) can be solved easily as compared with the CCSAREs (65) because there are no cross coupling terms. Specifically, although $\tilde{A}$ in CCSARE (65a) includes Ξ_i, $i = 1, \ldots, N$, $\hat{A}$ in SARE (62a) does not depend on other solutions, which means that CCSARE (65a) is a higher-order cross-coupled equation. Conversely, SARE (62a) is independent of other SAREs. Therefore, it is easy to solve N independent SARE (62a) because, when one derives Newton's method, it is easy to derive the difference equation. It should be noted it is quite difficult to derive Newton's algorithm to solve higher-order cross-coupled equations [42].

Finally, the design procedure is given below.

Step 1.   *Calculate the solution set $P^* > 0$ and $W^* > 0$ by solving the CCSAREs (59) to obtain the team-optimal control with H_∞ constraint that is described by (60).*

Step 2.   *By combining SAREs (62a) and MAEs (62b), compute Z_i^*, η_{ii}^* and calculate $\eta_{0i}^* = -R_{00i}^{-1} B_{0i}^T P^* + \eta_{ii}^* R_{0i}^{-1} B_i^T P^*$ for $i = 1, \ldots, N$.*

Step 3.   *Announce the incentive Stackelberg strategy $u_{0i}^*(t) = \eta_{0i}^* x(t) + \eta_{ii}^* u_i(t)$ to the ith follower, $i = 1, \ldots, N$, respectively.*

The flowchart of the above mentioned algorithm for calculating the incentive is shown in Fig. 4.

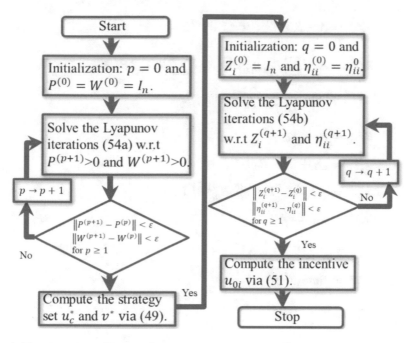

Fig. 4 Flowchart of algorithm for calculating the incentive

Although the CCSAREs (59) and the SAREs (62) seem to be very complicated, one can solve these matrix equations by using an iterative scheme. In order to solve the CCSAREs (59) and the SAREs (62), the following computational algorithms that are based on the Lyapunov iterations are given [43]:

$$
\begin{cases}
P^{(p+1)} A_c^{(p)} + A_c^{(p)T} P^{(p+1)} \\
\quad + A_p^T P^{(p+1)} A_p + P^{(p)} S_c P^{(p)} + Q_0 = 0, \\
W^{(p+1)} A_c^{(p)} + A_c^T W^{(p+1)} + A_p^T W^{(p+1)} A_p \\
\quad - \gamma^{-2} W^{(p)} T W^{(p)} + Q_0 + P^{(p)} S_c P^{(p)} = 0,
\end{cases}
\tag{66a}
$$

$$
\begin{cases}
Z_i^{(q+1)} \hat{A}^{(q)} + \hat{A}^{(q)T} Z_i^{(q+1)} + A_p^T Z_i^{(q+1)} A_p \\
\quad + Z_i^{(q)} \bar{B}_i^{(q)} [\hat{R}_{ii}^{(q)}]^{-1} \bar{B}_i^{(q)T} Z_i^{(q)} + \hat{Q}_i^{(q)} \\
\quad - \eta_{0i}^{(q)T} R_{0ii} \eta_{ii}^{(q)} [\hat{R}_{ii}^{(q)}]^{-1} \eta_{ii}^{(q)T} R_{0ii} \eta_{0i}^{(q)} = 0, \\
\eta_{ii}^{(q+1)T} (B_{0i}^T Z_i^{(q)} - R_{0ii} R_{00}^{-1} B_{0i}^T P) \\
\quad = R_{ii} R_{0i}^{-1} B_i^T P - B_i^T Z_i^{(q)},
\end{cases}
\tag{66b}
$$

where $i = 1, \ldots, N$, p, $q = 0, 1, \ldots$,

$$
P^{(0)} = Z_i^{(0)} = W^{(0)} = I_n, \quad \eta_{ii}^{(0)} = \eta_{ii}^0,
$$

$$A_c^{(p)} := A - S_c P^{(p)} + \gamma^{-2} T W^{(p)},$$

$$\hat{A}^{(q)} := A - \sum_{j=1}^{N} B_{0j} R_{00j}^{-1} B_{0j}^T P - \sum_{j=1}^{N} B_j R_{0j}^{-1} B_j^T P + \gamma^{-2} T W$$

$$\qquad - \bar{B}_i^{(q)} [\hat{R}_{ii}^{(q)}]^{-1} \bar{B}_i^{(q)T} Z_i^{(q)},$$

$$\bar{B}_i^{(q)} := B_i + B_{0i} \eta_{ii}^{(q)},$$

$$\hat{Q}_i^{(q)} := Q_i + \eta_{0i}^{(q)T} R_{0i} \eta_{0i}^{(q)},$$

$$\eta_{0i}^{(q)} := -R_{00i}^{-1} B_{0i}^T P + \eta_{ii}^{(q)} R_{0i}^{-1} B_i^T P,$$

$$\hat{R}_{ii}^{(q)} := R_{ii} + \eta_{ii}^{(q)T} R_{0i} \eta_{ii}^{(q)},$$

$$P = \lim_{p \to \infty} P^{(p)},$$

$$W = \lim_{p \to \infty} W^{(p)},$$

$$Z_i = \lim_{q \to \infty} Z_i^{(q)},$$

$$\eta_{ii} = \lim_{q \to \infty} \eta_{ii}^{(q)}.$$

It should be noted that the initial guess of $\eta_{ii}^{(0)} = \eta_{ii}^0$ has to be chosen appropriately. It should also be noted that the convergence rate of algorithm (66) is unclear. However, we will find from the numerical examples that this algorithm can work well in practice.

3.5 Numerical Example

In order to demonstrate the effectiveness of the incentive Stackelberg strategy set, a simple practical example in infinite-horizon case is investigated.

An R–L–C electrical circuit in Fig. 5 is considered [31]. In this network, R_i, r_i, $i = 1$, 2, R and L are the resistances and the inductance, respectively. The capacitances are denoted by C_i, $i = 1$, 2. Moreover, $E_{0i}(t)$ and $E_i(t)$, $i = 1$, 2 denote the applied voltages. $i(t)$ denotes the electric current in inductance L. In this survey, the leader has the main control inputs that represent the applied voltage, and the followers are the sub-controller corresponding to the parallel circuit, i.e., $E_{01}(t) := u_{01}(t)$, $E_{02}(t) := u_{02}(t)$ and $E_1(t) := u_1(t)$, $E_2(t) := u_2(t)$.

For this system, consider $L di(t)/dt = v(t)$ of the voltage drop across the inductor as an external deterministic disturbance. $V := x$, as a state, denotes the voltage drop across the circuit. It should be noted that, in any electronic

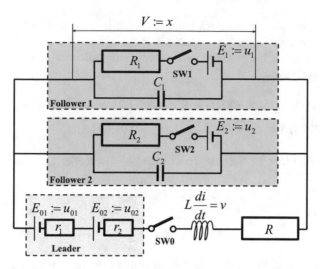

Fig. 5 Circuit diagram

device, thermal noise is unavoidable at non-zero temperatures. Thus, the system can be represented as a stochastic system governed by an Itô stochastic differential equation. If this noise is treated as a real-valued state-dependent Wiener process $w(t)$ with coefficient A_p, then the stochastic system can be written.

$$dx(t) = \left[Ax(t) + \sum_{i=1}^{2} [B_{0i}u_{0i}(t) + B_i u_i(t)] + Ev(t) \right] dt$$

$$+ A_p x(t) dw(t), \quad x(0) = x_0, \tag{67}$$

where

$$A = -\frac{1}{C_T} \left(\frac{1}{R_1} + \frac{1}{R_2} + \frac{1}{R_S} \right), \quad B_{01} = B_{02} = \frac{1}{C_T R_S},$$

$$B_1 = \frac{1}{C_T R_1}, \quad B_2 = \frac{1}{C_T R_2}, \quad E = -\frac{1}{C_T R_S}, \quad A_p = 0.01A,$$

$$C_T = C_1 + C_2, \quad R_S = R + r_1 + r_2.$$

It should be noted that system noise has been added to the deterministic system by describing it stochastically in the SDE (67). It is assumed that 1% of the magnitude of the state coefficient can be represented by a Wiener process based on stochastic perturbations. In this problem, it is assumed that the leader will control the voltage sources such that the team-optimal solution will be achieved, which will attenuate the external disturbance $v(t)$ under the H_∞ constraint. In contrast, with respect to the leader's incentive Stackelberg strategy, the followers will optimize their own costs simultaneously using Nash equilibrium strategies.

In order to solve this problem numerically, simulation data are assigned to the parameters as follows:

$$R_1 = 600 \ [\Omega], \ R_2 = 200 \ [\Omega], \ R = 1000 \ [\Omega], \ r_1 = 20 \ [\Omega], \ r_2 = 80 \ [\Omega],$$

$$C_1 = 8200 \ [\mu F], \ C_2 = 2200 \ [\mu F], \ L = 0.01 \ [H].$$

The weight matrices of the cost functionals of the leader and followers can be defined as

$$R_{001} = 2, \ R_{002} = 4, \ R_{01} = 3, \ R_{02} = 2, \ R_{011} = 4, \ R_{022} = 4,$$

$$R_{11} = 3, \ R_{22} = 2, \ Q_0 = 1, \ Q_1 = 2, \ Q_2 = 4.$$

Next, we select $\gamma = 2$ to design the incentive Stackelberg strategy set.

According the flowchart, the strategy set and incentive can be computed as follows:

$$u_c^*(t) = K_c^* x(t) = -R_c^{-1} B_c^T P^* x(t),$$

$$v^*(t) = F_\gamma^* x(t) = \gamma^{-2} E^T W^* x(t),$$

$$u_{0i}(t) = \eta_{0i}^* x(t) + \eta_{ii}^* u_i(t),$$

$$u_i^*(t) = K_i^* x(t) = u_i^\dagger(t) = -\hat{R}_{ii}^{-1} \left(\eta_{ii}^{*T} R_{0ii} \eta_{0i}^* + \bar{B}_i^T Z_i^* \right) x(t),$$

where $i = 1, \ 2$,

$$K_c^* = \begin{bmatrix} K_{01}^* \\ K_{02}^* \\ K_1^* \\ K_2^* \end{bmatrix} = \begin{bmatrix} -2.8406\text{e-}2 \\ -1.4203\text{e-}2 \\ -3.4719\text{e-}2 \\ -1.5624\text{e-}1 \end{bmatrix},$$

$$F_\gamma^* = 1.5000\text{e-}2,$$

$$P^* = 6.4994\text{e-}1,$$

$$W^* = 6.8641\text{e-}1,$$

$$\eta_{11}^* = 2.8782\text{e+}1,$$

$$\eta_{22}^* = -6.4766,$$

$$\eta_{01}^* = 9.7088\text{e-}1,$$

$$\eta_{02}^* = -1.0261,$$

$$Z_1^* = \Xi_1^* = 1.2350,$$

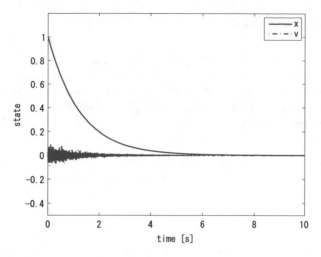

Fig. 6 Time history

$$Z_2^* = \varXi_2^* = 2.4801.$$

The result of the time history of this practical example is shown. The initial state is set to $x(0) = 1.0$.

Finally, the state variable $x(t) = V(t)$, the voltage drop across the circuit, and the disturbance $v(t) = Ldi(t)/dt$ are shown in Fig. 6. It can be observed that $v(t)$ is attenuated subject to the constraint boundary of the external deterministic disturbance under which the exponential decay of the capacitor's voltage over time can be obtained by using the proposed method.

If this technique is applied to the practical case, some drawbacks exist. For example, the full state information would not be obtained for the players. In this case, the observer based design or static output feedback strategy is reliable. As another approach, the decentralized technique seems to be standards. In this case, the appropriately cost functions and the constraint conditions can be considered.

4   Incentive Stackelberg Games for a Class of Markovian Jump Linear Stochastic Systems

In this section, an infinite-horizon incentive Stackelberg dynamic game for a class of continuous-time Markovian Jump Linear Stochastic Systems (MJLSSs) governed by Itô's differential equations is considered. It should be noted that in a two-level hierarchy, this game consists of multiple leaders and multiple followers.

4.1   Preliminary Results

Throughout this paper, we assume that $(\Omega, \mathcal{F}, \{\mathcal{F}_t\}_{t\geq 0}, P)$ is a given filtered probability space in which $w(t)$, $t \geq 0$ is a one-dimensional standard Wiener process, and r_t, $t \geq 0$ is a right continuous homogeneous Markov process with discrete state-space $\mathcal{D} = \{1, 2, \ldots, s\}$. We also assume that r_t is independent of $w(t)$ and has the probabilities of transition

$$P\{r_{t+h} = m \mid r_t = k\} = \begin{cases} \pi_{km}h + o(h), & \text{if } k \neq m, \\ 1 + \pi_{kk}h + o(h), & \text{if } k = m, \end{cases} \tag{68}$$

where

$$h > 0, \ \lim_{h\to 0} \frac{o(h)}{h} = 0, \ \pi_{km} \geq 0, \ k \neq m, \ \pi_{kk} = - \sum_{m=1, \ m\neq k}^{s} \pi_{km}.$$

The symbol π_{km} indicates the transition rate from modes k to m from times t to $t+h$ for all k, $m \in \mathcal{D}$. In order to establish the main results, the following definition and lemmas are required.

First, the H_2/H_∞ control problem in an infinite-horizon is introduced. Let us consider the following MJLSS with the deterministic disturbance $v(t)$:

$$dx(t) = [A(r_t)x(t) + B(r_t)u(t) + D(r_t)v(t)]dt$$

$$+ A_p(r_t)x(t)dw(t), \quad x(0) = x^0, \tag{69a}$$

$$z(t) = \begin{bmatrix} E(r_t)x(t) \\ u(t) \end{bmatrix}, \tag{69b}$$

where $x(t) \in \mathbb{R}^n$, $u(t) \in \mathbb{R}^m$, $v(t) \in \mathbb{R}^{n_v}$, $z(t) \in \mathbb{R}^{n_z}$ are the state, control input, external disturbance and controlled output, respectively. The coefficients $\mathbf{A}$, $\mathbf{A}_p \in \mathbb{M}_{n,n}^s$, $\mathbf{B} \in \mathbb{M}_{n,m}^s$, $\mathbf{D} \in \mathbb{M}_{n,n_v}^s$, $\mathbf{E} \in \mathbb{M}_{n_c,n}^s$ with $A(k)$, $B(k)$, $D(k)$, $A_p(k)$, and $E(k)$ are constant matrices of compatible dimensions, $k \in \mathcal{D}$.

Two associated performances can be defined as follows:

$$J_u(u, v; x^0, k) = \mathbb{E}\left[\int_0^\infty \|z(t)\|^2 dt \mid r_0 = k\right], \tag{70a}$$

$$J_\gamma(u, v; x^0, k) = \mathbb{E}\left[\int_0^\infty \left(\gamma^2 \|v(t)\|^2 - \|z(t)\|^2\right) dt \mid r_0 = k\right], \tag{70b}$$

where $\gamma > 0$ is a given disturbance attenuation level and it is the desired performance assigned prior by the control designer. The infinite-horizon stochastic H_2/H_∞ control of system (69) can be defined as follows [46]:

Definition 5 Suppose that the strategy pair $(u^*(t), v^*(t)) \in \mathcal{L}_{\mathcal{F}}^2(\mathbb{R}_+, \mathbb{R}^m) \times \mathcal{L}_{\mathcal{F}}^2(\mathbb{R}_+, \mathbb{R}^{n_v})$ exists such that

(i) when $v(t) = 0$, the closed-loop MJLSS (69) with some control $u^*(t)$ is asymptotically mean-square stable,

(ii) the H_∞-norm of the perturbed operator L_{u^*} satisfies the condition

$$\|L_{u^*}\|_\infty^2 = \sup_{\substack{v \in \mathcal{L}_{\mathcal{F}}^2(\mathbb{R}_+, \mathbb{R}^{n_v}), \\ v \neq 0, \, u = u^*, \, x^0 = 0}} \frac{\bar{J}_u(u^*; x^0)}{\bar{J}_v(u^*, v; x^0)} < \gamma^2, \tag{71}$$

where

$$\bar{J}_u(u^*; x^0) := \sum_{k=1}^s \mathbb{E}\left[\int_0^\infty \|z(t)\|^2 dt \mid r_0 = k\right],$$

$$\bar{J}_v(u^*, v; x^0) := \sum_{k=1}^s \mathbb{E}\left[\int_0^\infty \|v(t)\|^2 dt \mid r_0 = k\right].$$

(iii) when the worst-case disturbance $v^*(t) \in \mathcal{L}_{\mathcal{F}}^2(\mathbb{R}_+, \mathbb{R}^{n_v})$ is applied, $u^*(t) \in \mathcal{L}_{\mathcal{F}}^2(\mathbb{R}_+, \mathbb{R}^m)$ minimizes the cost functional (70a) with $J_u(u, v^*; x^0, k)$.

Then, a solution pair $(u^*(t), v^*(t))$ is called the infinite-horizon stochastic H_2/H_∞ control strategy pair.

Clearly, the solution pair $(u^*(t), v^*(t))$ seems to be satisfying the Nash equilibrium properties

$$J_u(u^*, v^*; x^0, k) \leq J_u(u, v^*; x^0, k), \quad k \in \mathcal{D}, \tag{72a}$$

$$0 \leq J_\gamma(u^*, v^*; x^0, k) \leq J_\gamma(u^*, v; x^0, k), \quad k \in \mathcal{D}. \tag{72b}$$

If the admissible control $u^*(t) \in \mathcal{L}_{\mathcal{F}}^2(\mathbb{R}_+, \mathbb{R}^m)$ satisfies conditions (i) and (ii), then the infinite-horizon H_2/H_∞ control problem is termed an H_∞ control problem. In this case, consider that the state-feedback optimal control $u^*(t)$ minimizes the following cost functional:

$$J_u(u, v^*; x^0, k) = \mathbb{E}\left[\int_0^\infty \{x^T(t) Q(r_t) x(t) + u^T(t) R(r_t) u(t)\} dt \mid r_0 = k\right], \tag{73}$$

where $Q(k) = Q^T(k) \geq 0$ and $R(k) = R^T(k) > 0$ are constant matrices of compatible dimensions, $k \in \mathcal{D}$. Then Theorem 3.1 in [46] can be written in a simplified form adopting current notations through the following lemma for easy application in the main content.

Lemma 3 ([46]) *For MJLSS (69) associated with cost functionals (70), suppose that there exist solutions* $\mathbf{P} \geq \mathbf{0}$ *and* $\mathbf{W} \geq \mathbf{0}$ *to the following coupled SAREs:*

$$P(k)A_C(k) + A_E^T(k)P(k) + A_p^T(k)P(k)A_p(k) + \sum_{m=1}^{s} \pi_{km} P(m)$$

$$+ P(k)B(k)R^{-1}(k)B^T(k)P(k) + Q(k) = 0, \tag{74a}$$

$$W(k)A_C(k) + A_E^T(k)W(k) + A_p^T(k)W(k)A_p(k)$$

$$+ \sum_{m=1}^{s} \pi_{km} W(m) - \gamma^{-2}W(k)D(k)E^T(k)W(k)$$

$$+ P(k)B(k)R^{-2}(k)B^T(k)P(k) + C^T(k)E(k) = 0, \tag{74b}$$

where

$$A_C(k) := A(k) + B(k)K(k) + D(k)K_\gamma(k),$$

$$K(k) := -R^{-1}(k)B^T(k)P(k),$$

$$K_\gamma(k) := \gamma^{-2}E^T(k)W(k).$$

If $(\mathbf{A}, \ \mathbf{A}_p \mid \mathbf{C})$ *and* $(\mathbf{A} + \gamma^{-2}\mathbf{E}\mathbf{E}^T\mathbf{W}, \ \mathbf{A}_p \mid \mathbf{C})$ *are stochastically detectable, then the* H_∞ *constrained disturbance attenuation problem has the following pair of solutions satisfying the conditions of Definition 2,*

$$u^*(t) = K(r_t)x(t) = \sum_{k=1}^{s} K(k)x(t)\chi_{r_t=k}, \tag{75a}$$

$$v^*(t) = K_\gamma(r_t)x(t) = \sum_{k=1}^{s} K_\gamma(k)x(t)\chi_{r_t=k}. \tag{75b}$$

Second, let us introduce the stochastic LQ control with a jump in the following form:

$$\min \ \mathbb{E}\left[\int_0^\infty \{x^T(t)Q(r_t)x(t) + 2x^T(t)L(r_t)u(t)\right.$$

$$\left. + u^T(t)R(r_t)u(t)\}dt \mid r_0 = k\right], \tag{76a}$$

$$\text{s.t. } dx(t) = [A(r_t)x(t) + B(r_t)u(t)]dt$$

$$+ A_p(r_t)x(t)dw(t), \quad x(0) = x^0, \tag{76b}$$

where the coefficients $Q(k) = Q^T(k) \geq 0$, $L(k)$ and $R(k) = R^T(k) > 0$ are constant matrices of compatible dimensions, $k \in \mathcal{D}$. In this case, the following result is known [32, 33].

Lemma 4 ([32, 33]) *For LQ problem (76), suppose that there exists a symmetric solution* $\mathbf{X} \geq \mathbf{0}$ *to the following SAREs:*

$$X(k)A(k) + A^T(k)X(k) + A_p^T(k)X(k)A_p(k)$$

$$+ \sum_{m=1}^{s} \pi_{km} X(m) - [X(k)B(k) + L(k)]R^{-1}(k)$$

$$\times [X(k)B(k) + L(k)]^T + Q(k) = 0, \quad \forall k \in \mathcal{D}. \tag{77}$$

If $\mathbf{Q} - \mathbf{LR}^{-1}\mathbf{L}^T \geq 0$ *and* $\left(\mathbf{A}, \mathbf{A}_p \mid \sqrt{\mathbf{Q} - \mathbf{LR}^{-1}\mathbf{L}^T}\right)$ *is stochastically detectable, then the optimal feedback control of LQ problem (76) is*

$$u^*(t) = -\sum_{k=1}^{s} R^{-1}(k)[X(k)B(k) + L(k)]^T x(t)\chi_{r_t=k}. \tag{78}$$

4.2 Problem Formulation

Consider the following MJLSS with state-dependent noise:

$$dx(t) = \left[A(r_t)x(t) + \sum_{i=1}^{M}\sum_{j=1}^{N}\left[B_{Lij}(r_t)u_{Lij}(t) + B_{Fji}(r_t)u_{Fji}(t)\right]\right.$$

$$\left. + D(r_t)v(t)\right]dt + A_p(r_t)x(t)dw(t), \quad x(0) = x^0, \tag{79a}$$

$$z(t) = \text{col}\left[E(r_t)x(t)\ u_{c1}(t)\ u_{c2}(t)\ \cdots\ u_{cM}(t)\right], \tag{79b}$$

where

$$u_{ci}(t) = \text{col}\left[u_{Li1}(t)\ \cdots\ u_{LiN}(t)\ u_{F1i}(t)\ \cdots\ u_{FNi}(t)\right].$$

$u_{Lij}(t) \in \mathbb{R}^{n_{Lij}}$ represents the leader L_i's control input for the follower F_j and $u_{Fji}(t) \in \mathbb{R}^{n_{Fji}}$ represents the follower F_j's control input according to the leader L_i in the sense of incentive Stackelberg strategy; the coefficients $\mathbf{A}$, $\mathbf{B}_{Lij}$, $\mathbf{B}_{Fji}$, $\mathbf{E}$, $\mathbf{A}_p$, and $\mathbf{C}$ are constant matrices of compatible dimensions, $i = 1, \ldots, M$, $j = 1, \ldots, N$.

Cost functionals of the leader L_i and the follower F_j are accordingly given by

$$J_{Li}\left(u_{Li1}, \ldots, u_{LiN}, u_{F1i}, \ldots, u_{FNi}, v; x^0, k\right)$$

$$= \mathbb{E}\left[\int_0^\infty \left\{x^T(t)Q_{Li}(r_t)x(t) + \sum_{j=1}^N \left[u_{Lij}^T(t)R_{Lij}(r_t)u_{Lij}(t)\right.\right.\right.$$

$$\left.\left.\left. + u_{Fji}^T(t)R_{LFji}(r_t)u_{Fji}(t)\right]\right\}dt \mid r_0 = k\right],\tag{80a}$$

$$J_{Fj}\left(u_{L1j}, \ldots, u_{LMj}, u_{Fj1}, \ldots, u_{FjM}, v; x^0, k\right)$$

$$= \mathbb{E}\left[\int_0^\infty \left\{x^T(t)Q_{Fj}(r_t)x(t) + \sum_{i=1}^M \left[u_{Lij}^T(t)R_{FLij}(r_t)u_{Lij}(t)\right.\right.\right.$$

$$\left.\left.\left. + u_{Fji}^T(t)R_{Fji}(r_t)u_{Fji}(t)\right]\right\}dt \mid r_0 = k\right],\tag{80b}$$

where $Q_{Li}(k) = Q_{Li}^T(k) \geq 0$, $Q_{Fj}(k) = Q_{Fj}^T(k) \geq 0$, $R_{Lij}(k)$
$= R_{Lij}^T(k) > 0$, $R_{Fji} = R_{Fji}^T(k) > 0$, $R_{LFji}(k) = R_{LFji}^T(k)$
≥ 0 and $R_{FLij}(k) = R_{FLij}^T(k) \geq 0$, $k \in \mathcal{D}$, $i = 1, \ldots, M$, $j = 1, \ldots, N$.
For an incentive Stackelberg game, leaders announce the following incentive
strategy to the followers ahead of time:

$$u_{Lij}(t) = \Theta_{Lij}(r_t)x(t) + \Theta_{Fji}(r_t)u_{Fji}(t)$$

$$= \sum_{k=1}^s \Theta_{Lij}(k)x(t)\chi_{r_t=k} + \Theta_{Fji}(r_t)u_{Fji}(t),\tag{81}$$

where the parameters $\Theta_{Lij}(k)$ and $\Theta_{Fji}(k)$ are to be determined as associated
with the Nash equilibrium or Pareto optimal strategies $u_{Fji}(t)$ of the followers for
$k \in \mathcal{D}$, $i = 1, \ldots, M$, $j = 1, \ldots, N$. In this game, leaders will achieve a
Nash equilibrium or Pareto optimal solution attenuating the external disturbance
$v(t)$ with an H_∞ constraint. Infinite-horizon multiple leader-follower incentive
Stackelberg games for MJLSSs with an H_∞ constraint can be formulated as follows:
As introduced in [32], in an practical system, a hierarchical structure having a large
number of readers and a large number of followers is common.

For a given disturbance attenuation level $\gamma > 0$, find, if possible, the state
feedback controls

$$u_{Lij}^*(t) = K_{Lij}^*(r_t)x(t) = \sum_{k=1}^s K_{Lij}^*(k)x(t)\chi_{r_t=k},\tag{82a}$$

$$u^*_{Fji}(t) = K^*_{Fji}(r_t)x(t) = \sum_{k=1}^{s} K^*_{Fji}(k)x(t)\chi_{r_t=k} \tag{82b}$$

such that the following hold:

(i) The trajectory of MJLSS (79) satisfies the Nash equilibrium conditions (83a) of the leaders with an H_∞ constraint condition (83b):

$$J_{Li}(u^*_{c1}, \ldots, u^*_{cM}, v^*; x^0, k) \leq J_{Li}(\gamma^*_{-i}(u_{ci}), v^*; x^0, k), \tag{83a}$$

$$0 \leq J_\gamma(u^*_{c1}, \ldots, u^*_{cM}, v^*; x^0, k) \leq J_\gamma(u^*_{c1}, \ldots, u^*_{cM}, v; x^0, k), \tag{83b}$$

where $i = 1, \ldots, M$ and

$$J_\gamma(u_{c1}, \ldots, u_{cM}, v; x^0, k)$$

$$= \mathbb{E}\left[\int_0^\infty \left\{ \gamma^2\|v(t)\|^2 - \|z(t)\|^2 \right\}dt \;\Big|\; r_0 = k \right], \; v(t) \neq 0,$$

$$\|z(t)\|^2 = x^T(t)C^T(r_t)E(r_t)x(t) + \sum_{i=1}^{M} u^T_{ci}(t)u_{ci}(t).$$

Consider the leader's incentive strategy (81) and the worst-case disturbance $v^*(t) \in \mathcal{L}^2_{\mathcal{F}}(\mathbb{R}_+, \mathbb{R}^{n_v})$. The follower's decision $u^*_{Fji}(t) \in \mathcal{L}^2_{\mathcal{F}}(\mathbb{R}_+, \mathbb{R}^{n_{Fji}})$, $i = 1, \ldots, M$ can be selected as follows:

(ii-a) If Nash equilibrium as non-cooperative strategy is chosen, find the Nash strategy set such that the following inequality holds:

$$J_{Fj}(\hat{u}^*_{F1}, \ldots, \hat{u}^*_{FN}, v^*; x^0, k) \leq J_{Fj}(\gamma^*_{-j}(\hat{u}_{Fj})), v^*; x^0, k), \tag{84}$$

where

$$\hat{u}_{Fj}(t) = \mathbf{col}\left[u_{Fj1}(t) \cdots u_{FjM}(t)\right], \; j = 1, \ldots, N.$$

(ii-b) If the Pareto optimal strategy as cooperative strategy is chosen, the following objective function should be optimized.

$$\hat{J}_\rho(\hat{u}_{F1}, \ldots, \hat{u}_{FN}, v; x^0, k) = \sum_{j=1}^{N} \rho_j J_{Fj}(\hat{u}_{F1}, \ldots, \hat{u}_{FN}, v; x^0, k), \tag{85}$$

where $\sum_{j=1}^{N} \rho_j = 1, 0 < \rho_j < 1, j = 1, \ldots, N.$

4.3 Main Results

In this section, the leader's Nash equilibrium strategy and the follower's strategy sets are derived.

4.4 Leader's Nash Equilibrium Strategy

It is assumed that leaders are regarded as non-cooperative in their group, and they use Nash equilibrium. Thus, a leader's Nash equilibrium solutions $(u_{c1}^*(t), \ldots, u_{cM}^*(t), v^*(t))$ are investigated attenuating the disturbance under an H_∞ constraint. For this purpose, let us configure the MJLSS (79) into the following centralized system:

$$dx(t) = \left[A(r_t)x(t) + \sum_{i=1}^{M} B_{ci}(r_t)u_{ci}(t) + D(r_t)v(t) \right] dt$$

$$+ A_p(r_t)x(t)dw(t), \quad x(0) = x^0, \tag{86a}$$

$$z(t) = \mathbf{col}\left[E(r_t)x(t) \ u_{c1}(t) \ u_{c2}(t) \cdots u_{cM}(t) \right], \tag{86b}$$

where

$$B_{ci}(k) = \left[B_{Li1}(k) \cdots B_{LiN}(k) \ B_{F1i}(k) \cdots B_{FNi}(k) \right],$$

$i = 1, \ldots, M$ and $k \in \mathcal{D}$.

Furthermore, the cost functional (80a) can be changed as follows:

$$J_{Li}\left(u_{ci}; \ x^0, \ k \right) = \mathbb{E}\left[\int_0^\infty \left\{ x^T(t)Q_{Li}(r_t)x(t) \right. \right.$$

$$\left. \left. + u_{ci}^T(t)R_{ci}(r_t)u_{ci}(t) \right\} dt \middle| r_0 = k \right], \tag{87}$$

where

$$R_{ci}(k) = \mathbf{block\ diag}\left(R_{Li1}(k) \cdots R_{LiN}(k) \ R_{LF1i}(k) \cdots R_{LFNi}(k) \right).$$

In order to obtain the ith leader's Nash equilibrium strategies with H_∞ disturbance control, the following result can be derived through Lemma 3.

Corollary 1 *For a given disturbance attenuation level $\gamma > 0$, suppose that the following coupled SAREs have solutions $\mathbf{P}_{ci} > 0$, $\mathbf{W} > 0$.*

$$P_{ci}(k)A_C(k) + A_E^T(k)P_{ci}(k) + A_p^T(k)P_{ci}(k)A_p(k)$$

$$+ \sum_{m=1}^{s} \pi_{km} P_{ci}(m) + P_{ci}(k)S_{ci}(k)P_{ci}(k) + Q_{Li}(k) = 0, \qquad (88a)$$

$$W(k)A_C(k) + A_E^T(k)W(k) + A_p^T(k)W(k)A_p(k)$$

$$+ \sum_{m=1}^{s} \pi_{km} W(m) - \gamma^{-2}W(k)T(k)W(k) + Q_C(k) = 0, \qquad (88b)$$

where $i = 1, \ldots, M$

$$A_C(k) := A(k) - \sum_{i=1}^{M} S_{ci}(k)P_{ci}(k) + \gamma^{-2}T(k)W(k),$$

$$S_{ci}(k) := B_{ci}(k)R_{ci}^{-1}(k)B_{ci}^T(k),$$

$$T(k) := D(k)E^T(k),$$

$$Q_C(k) := \sum_{i=1}^{M} P_{ci}(k)B_{ci}(k)R_{ci}^{-2}(k)B_{ci}^T(k)P_{ci}(k) + C^T(k)E(k).$$

It is assumed that $(\mathbf{A}, \mathbf{A}_p \mid \mathbf{C})$ *and* $(\mathbf{A} + \gamma^{-2}\mathbf{E}\mathbf{E}^T\mathbf{W}, \mathbf{A}_p \mid \mathbf{C})$ *are stochastically detectable. Then, the* H_∞ *constrained disturbance attenuation problem has the following solutions:*

$$u_{ci}^*(t) = K_{ci}^*(r_t)x(t) = \sum_{k=1}^{s} K_{ci}^*(k)x(t)\chi_{r_t=k}, \qquad (89a)$$

$$v^*(t) = K_\gamma^*(r_t)x(t) = \sum_{k=1}^{s} K_\gamma^*(k)x(t)\chi_{r_t=k}, \qquad (89b)$$

where

$$K_{ci}^*(k) = \begin{bmatrix} K_{Li1}^*(k) \\ \vdots \\ K_{LiN}^*(k) \\ K_{F1i}^*(k) \\ \vdots \\ K_{FNi}^*(k) \end{bmatrix} = \begin{bmatrix} -R_{Li1}^{-1}(k)B_{Li1}^T(k)P_{ci}(k) \\ \vdots \\ -R_{LiN}^{-1}(k)B_{LiN}^T(k)P_{ci}(k) \\ -R_{LF1i}^{-1}(k)B_{F1i}^T(k)P_{ci}(k) \\ \vdots \\ -R_{LFNi}^{-1}(k)B_{FNi}^T(k)P_{ci}(k) \end{bmatrix},$$

$$K_\gamma^*(k) = \gamma^{-2}E^T(k)W(k), \quad k \in \mathcal{D}, \ i = 1, \ldots, M.$$

It should be noted that the relation between $\Theta_{Lij}(k)$ and $\Theta_{Fij}(k)$, $k \in \mathcal{D}$, $i = 1, \ldots, M$, $j = 1, \ldots, N$ can be derived from (81) as

$$\Theta_{Lji}(k) = -R_{Lij}^{-1}(k)B_{Lij}^T(k)P_{ci}(k)$$
$$+ \Theta_{Fji}(k)R_{LFji}^{-1}(k)B_{Fji}^T(k)P_{ci}(k). \qquad (90)$$

Using this relationship (90), the leader's incentive strategy (81) can be written in the following reduced parameter form:

$$u_{Lij}(t) = -R_{Lij}^{-1}(r_t)B_{Lij}^T(r_t)P_{ci}(r_t)x(t) + \Theta_{Fji}(r_t)$$
$$\times \left[u_{Fji}(t) + R_{LFji}^{-1}(r_t)B_{Fji}^T(r_t)P_{ci}(r_t)x(t)\right], \qquad (91)$$

for $i = 1, \ldots, M$, $j = 1, \ldots, N$. In order to find the leader's incentive strategy (91), we need to determine parameter $\Theta_{Fji}(k)$, $k \in \mathcal{D}$. For this purpose, optimization problems for the followers should be considered.

4.5   Follower's Nash Equilibrium Strategy

When the followers do not cooperate with each other, the followers' Nash equilibrium is applied based on the leader's incentive strategy (81) or (91) and the worst-case disturbance, $v^*(t)$. Centralizing (79a) with respect to the follower's inputs, the following stochastic system can be found.

$$dx(t) = \left[\hat{A}_{-j}(r_t)x(t) + B_{\theta j}(r_t)\hat{u}_{Fj}(t)\right]dt$$
$$+ A_p(r_t)x(t)dw(t), \qquad (92)$$

where

$$\hat{u}_{Fj}(t) = \mathbf{col}\left[u_{Fj1}(t) \cdots u_{FjM}(t)\right],$$

$$\hat{A}_{-j}(k) := A(k) + \sum_{i=1}^{M}\sum_{j=1}^{N} B_{Lij}(k)\Theta_{Lij}(k)$$

$$+ \sum_{\ell=1, \ell\neq j}^{N} B_{\theta\ell}(k)K_{F\ell}^*(k) + \gamma^{-2}T(k)W(k),$$

$$B_{\theta j}(k) := \left[B_{\theta j1}(k) \cdots B_{\theta jM}(k)\right],$$

$$B_{\theta ji}(k) := B_{Fji}(k) + B_{Lij}(k)\Theta_{Fji}(k).$$

The cost functional (80b) of the jth follower can be rewritten as

$$J_{Fj}\big(u_{L1j}(\hat{u}_{Fj}), \ldots, u_{LMj}(\hat{u}_{Fj}), \hat{u}_{Fj}, v^*; x^0, k\big)$$

$$= \mathbb{E}\bigg[\int_0^\infty \bigg\{x^T(t)\hat{Q}_{Fj}(r_t)x(t) + 2x^T(t)S_{\theta j}(r_t)\hat{u}_{Fj}(t)$$

$$+ \hat{u}_{Fj}^T(t)R_{\theta j}(r_t)\hat{u}_{Fj}(t)\bigg\}dt\bigg| r_0 = k\bigg], \tag{93}$$

where

$$\hat{Q}_{Fj}(k) := Q_{Fj}(k) + \sum_{i=1}^M \Theta_{Lji}^T(k)R_{FLij}(k)\Theta_{Lji}(k),$$

$$R_{\theta j}(k) := \textbf{block diag}\big(R_{\theta j1}(k) \cdots R_{\theta jM}(k) \big),$$

$$R_{\theta ji}(k) := R_{Fji}(k) + \Theta_{Fji}^T(k)R_{FLij}(k)\Theta_{Fji}(k),$$

$$S_{\theta j}(k) := \big[S_{\theta j1}(k) \cdots S_{\theta jM}(k) \big],$$

$$S_{\theta ji}(k) := \Theta_{Lji}^T(k)R_{FLij}(k)\Theta_{Fji}(k).$$

Hence, by applying Lemma 4, the jth follower's Nash equilibrium strategy can be obtained through the following SAREs.

$$P_{Fj}(k)\hat{A}_{-j}(k) + \hat{A}_{-j}^T(k)P_{Fj}(k) + A_p^T(k)P_{Fj}(k)A_p(k)$$

$$+ \sum_{m=1}^s \pi_{km} P_{Fj}(m) - [P_{Fj}(k)B_{\theta j}(k) + S_{\theta j}(k)]R_{\theta j}^{-1}(k)$$

$$\times [P_{Fj}(k)B_{\theta j}(k) + S_{\theta j}(k)]^T + \hat{Q}_{Fj}(k) = 0. \tag{94}$$

If $\hat{Q}_{Fj}(k) - S_{\theta j}(k)R_{\theta j}^{-1}(k)S_{\theta j}^T(k) \geq 0$ and $\big(\hat{A}_{-j}, A_p \mid \sqrt{\hat{Q}_{Fj} - \hat{S}_{Fj}\hat{R}_{Fj}^{-1}\hat{S}_{Fj}^T}\big)$ is stochastically detectable, then the follower's Nash equilibrium strategy set is given by

$$\hat{u}_{Fj}(t) = \hat{K}_{Fj}(r_t)x(t) = \sum_{k=1}^s \hat{K}_{Fj}(k)x(t)\chi_{r_t=k}, \tag{95}$$

where $j = 1, \ldots, N$,

$$\hat{K}_{Fj}(k) := -R_{\theta j}^{-1}(k)\big[B_{\theta j}^T(k)P_{Fj}(k) + S_{\theta j}^T(k) \big]$$

$$= \textbf{col}\big[\hat{K}_{Fj1}(k) \cdots \hat{K}_{FjM}(k) \big],$$

$$\hat{K}_{Fji}(k) := -R_{\theta ji}^{-1}(k)\left[B_{\theta ji}^T(k)P_{Fj}(k) + S_{\theta ji}^T(k)\right].$$

In addition, we can find the follower's matrix gain from the leader's centralized matrix gain (89a) as follows:

$$K_{Fji}^*(k) = -R_{LFji}^{-1}(k)B_{Fji}^T(k)P_{ci}(k). \tag{96}$$

From (95) and (96), we can establish the equivalence relation $K_{Fji}^*(k) \equiv \hat{K}_{Fji}(k)$ through the following MAEs:

$$- R_{\theta ji}(k)K_{Fji}^*(k) = B_{\theta ji}^T(k)P_{Fj}(k) + S_{\theta ji}^T(k),$$

$$k \in \mathcal{D}, \ i = 1,\ldots,M, \ j = 1,\ldots,N. \tag{97}$$

From the MAEs (97), $\Theta_{Fji}(k)$ can be computed through (90). Finally, the following result can be obtained.

Theorem 3 ([33]) *Suppose that the coupled SAREs (88), SAREs (94), and MAEs (97) have solutions. Then, incentive (91) associated with (89) and (97) constitutes an incentive Stackelberg strategy set with an H_∞ constraint when the followers act in a non-cooperative.*

Remark 2 Substituting $\Theta_{Lij}(k)$ from relation (90) into the SAREs (94), two unknown matrices $P_{Fj}(k)$ and $\Theta_{Fji}(k)$ can be obtained by solving SAREs (94) and MAEs (97).

4.6 Follower's Pareto Optimal Strategy

When the followers act cooperatively, the follower's Pareto optimal strategy is calculated based on the leader's incentive strategy (81) or (91) and the worst-case disturbance $v^*(t)$. Centralizing (79a) with respect to the follower's inputs, the following system can be found.

$$dx(t) = \left[\tilde{A}(r_t)x(t) + B_\theta(r_t)\tilde{u}_F(t)\right]dt + A_p(r_t)x(t)dw(t), \tag{98}$$

where

$$\tilde{u}_F(t) = \mathbf{col}\left[\tilde{u}_{F1}(t) \cdots \tilde{u}_{FN}(t)\right],$$

$$\tilde{u}_{Fj}(t) = \mathbf{col}\left[u_{Fj1}(t) \cdots u_{FjM}(t)\right],$$

$$\tilde{A}(k) := A(k) + \sum_{i=1}^{M} \sum_{j=1}^{N} B_{Lij}(k)\Theta_{Lij}(k) + \gamma^{-2}T(k)W(k),$$

$$B_\theta(k) := [B_{\theta 1}(k) \ \cdots \ B_{\theta N}(k)].$$

In order to find the followers' Pareto optimal strategy, we have the following cost functional from (80b).

$$J_\rho(\tilde{u}_F, v^*; x^0, k) = \sum_{j=1}^{N} \rho_j J_{Fj}\left(\tilde{u}_{Fj}, v^*; x^0, k\right)$$

$$= \mathbb{E}\left[\int_0^\infty \left\{x^T(t)\tilde{Q}(r_t)x(t) + 2x^T(t)\tilde{S}(r_t)\tilde{u}_F(t)\right.\right.$$

$$\left.\left. + \tilde{u}_F^T(t)\tilde{R}(r_t)\tilde{u}_F(t)\right\}dt \,\middle|\, r_0 = k\right], \tag{99}$$

where

$$\tilde{Q}(k) := \sum_{j=1}^{N} \rho_j \hat{Q}_{Fj}(k),$$

$$\tilde{S}(k) := [\rho_1 S_{\theta 1}(k) \ \cdots \ \rho_N S_{\theta N}(k)],$$

$$\tilde{R}(k) := \textbf{block diag}\big(\rho_1 R_{\theta 1}(k) \ \cdots \ \rho_N R_{\theta N}(k)\big).$$

Hence, by applying Lemma 4, the Pareto optimal strategy of the followers can be obtained through the following SAREs.

$$P_F(k)\tilde{A}(k) + \tilde{A}^T(k)P_F(k) + A_p^T(k)P_F(k)A_p(k)$$

$$+ \sum_{m=1}^{s} \pi_{km} P_F(m) - [P_F(k)B_\theta(k) + \tilde{S}(k)]\tilde{R}^{-1}(k)$$

$$\times [P_F(k)B_\theta(k) + \tilde{S}(k)]^T + \tilde{Q}(k) = 0. \tag{100}$$

If $\tilde{Q}(k) - \tilde{S}(k)\tilde{R}^{-1}(k)\tilde{S}^T(k) \geq 0$ and $\left(\tilde{\mathbf{A}}, \mathbf{A}_p \,\middle|\, \sqrt{\tilde{\mathbf{Q}} - \tilde{\mathbf{S}}\tilde{\mathbf{R}}^{-1}\tilde{\mathbf{S}}^T}\right)$ is stochastically detectable, then the Pareto optimal state feedback control is given by

$$\tilde{u}_{Fj}(t) = \tilde{K}_{Fj}(r_t)x(t) = \sum_{k=1}^{s} \tilde{K}_{Fj}(k)x(t)\chi_{r_t=k}, \tag{101}$$

where

$$\tilde{K}_{Fj}(k) = -\rho_j^{-1} R_{\theta j}^{-1}(k) \left[B_{\theta j}^T(k) P_F(k) + \rho_j S_{\theta j}^T(k) \right]$$

$$= \mathbf{col} \left[\tilde{K}_{Fj1}(k) \cdots \tilde{K}_{FjM}(k) \right],$$

$$\tilde{K}_{Fji}(k) = -\rho_j^{-1} R_{\theta ji}^{-1}(k) \left[B_{\theta ji}^T(k) P_F(k) + \rho_j S_{\theta ji}^T(k) \right].$$

Using the similar step of the Nash games for the previous subsection, we can establish the equivalence relation $K_{Fji}^*(k) \equiv \tilde{K}_{Fji}(k)$ through the following MAEs:

$$- \rho_j R_{\theta ji}(k) K_{Fji}^*(k) = B_{\theta ji}^T(k) P_F(k) + \rho_j S_{\theta ji}^T(k),$$

$$k \in \mathcal{D}, \ i = 1, \ldots, M, \ j = 1, \ldots, N. \tag{102}$$

From the MAEs (102), $\Theta_{Fji}(k)$ can also be computed. We are now in a position to state other result under the cooperative strategy for the followers.

Theorem 4 ([33]) *Suppose that the coupled SAREs (88), SAREs (100), and MAEs (102) have solutions. Then, incentive (91) associated with (89) and (102) constitutes an incentive Stackelberg strategy with the H_∞ constraint when the followers act in a cooperatively.*

Remark 3 The coupled SAREs (88) can be solved using several numerical methods such as a linear matrix inequality technique [47], Lyapunov iterations [27], Newton's method [44]. Furthermore, the SAREs (94) with the MAEs (97) or the SAREs (100) with the MAEs (102) can also be solved using the Lyapunov iterations.

4.7   Numerical Example

In order to demonstrate the efficiency of the proposed strategies numerically, let us consider an MJLSS with two modes. In this example, two leaders and two followers are considered and it is assumed that the followers choose the Nash equilibrium. The matrices of the MJLSS are given below.

$$M = 2, \ N = 2, \ s = 2, \ \pi = \begin{bmatrix} -0.2 & 0.2 \\ 0.8 & -0.8 \end{bmatrix},$$

$$Q_{L1}(k) = 2.5 I_2, \ Q_{L2}(k) = 2.2 I_2,$$

$$Q_{F1}(k) = 1.5 I_2, \ Q_{F1}(k) = 3.0 I_2, \ k = 1, \ 2,$$

$$A(1) = \begin{bmatrix} 0.9 & 0 \\ 1.2 & -2.9 \end{bmatrix}, \ A_p(1) = \begin{bmatrix} 0 & 0.2 \\ 0.1 & 0 \end{bmatrix},$$

$$B_{L11}(1) = \begin{bmatrix} 0.13 & 0.20 \\ -0.55 & 0.81 \end{bmatrix}, \quad B_{L12}(1) = \begin{bmatrix} 0.31 & 1.20 \\ -1.25 & 1.02 \end{bmatrix},$$

$$B_{L21}(1) = \begin{bmatrix} 0.28 & 0.12 \\ 5.32 & 0 \end{bmatrix}, \quad B_{L22}(1) = \begin{bmatrix} 0.12 & 0.56 \\ 1.02 & 0.32 \end{bmatrix},$$

$$B_{F11}(1) = \begin{bmatrix} 0.15 & -0.11 \\ 0.55 & 1.32 \end{bmatrix}, \quad B_{F12}(1) = \begin{bmatrix} 0.51 & 0.54 \\ 0.21 & 1.21 \end{bmatrix},$$

$$B_{F21}(1) = \begin{bmatrix} 0.23 & -0.45 \\ 0.28 & 2.96 \end{bmatrix}, \quad B_{F22}(1) = \begin{bmatrix} 0.21 & 0.21 \\ 2.11 & 1.86 \end{bmatrix},$$

$$D(1) = \begin{bmatrix} 0.54 & 0.43 \\ 0.23 & 0.13 \end{bmatrix}, \quad E(1) = \begin{bmatrix} 1 & 2 \end{bmatrix},$$

$$R_{L11}(1) = 2.0I_2, \ R_{L12}(1) = 1.5I_2, \ R_{L21}(1) = 0.5I_2,$$

$$R_{L22}(1) = 2.2I_2, \ R_{F11}(1) = 3.5I_2, \ R_{F12}(1) = I_2,$$

$$R_{F21}(1) = 0.2I_2, \ R_{F22}(1) = 3.0I_2,$$

$$R_{LF11}(1) = \mathbf{diag}\begin{bmatrix} 3.0 & 2.0 \end{bmatrix}, \ R_{LF12}(1) = \mathbf{diag}\begin{bmatrix} 1.0 & 1.5 \end{bmatrix},$$

$$R_{LF21}(1) = \mathbf{diag}\begin{bmatrix} 2.0 & 1.2 \end{bmatrix}, \ R_{LF22}(1) = \mathbf{diag}\begin{bmatrix} 1.2 & 1.5 \end{bmatrix},$$

$$R_{FL11}(1) = \mathbf{diag}\begin{bmatrix} 1.0 & 1.5 \end{bmatrix}, \ R_{FL12}(1) = \mathbf{diag}\begin{bmatrix} 4.0 & 2.0 \end{bmatrix},$$

$$R_{FL21}(1) = \mathbf{diag}\begin{bmatrix} 3.0 & 1.6 \end{bmatrix}, \ R_{FL22}(1) = \mathbf{diag}\begin{bmatrix} 2.5 & 1.9 \end{bmatrix},$$

$$A(2) = \begin{bmatrix} -1 & 0.2 \\ -0.5 & 1.5 \end{bmatrix}, \quad A_p(2) = \begin{bmatrix} 0.8 & 0 \\ 0.2 & 0 \end{bmatrix},$$

$$B_{L11}(2) = \begin{bmatrix} -0.32 & 0.12 \\ 1.23 & -0.92 \end{bmatrix}, \quad B_{L12}(2) = \begin{bmatrix} 0.13 & 3.11 \\ 0.53 & -1.21 \end{bmatrix},$$

$$B_{L21}(2) = \begin{bmatrix} -0.81 & 0.28 \\ 2.23 & -2.82 \end{bmatrix}, \quad B_{L22}(2) = \begin{bmatrix} 0.28 & 1.61 \\ -3.12 & 0.22 \end{bmatrix},$$

$$B_{F11}(2) = \begin{bmatrix} 0.52 & 0.52 \\ -0.51 & 1.22 \end{bmatrix}, \quad B_{F12}(2) = \begin{bmatrix} 0.18 & -1.41 \\ 0.18 & 2.10 \end{bmatrix},$$

$$B_{F21}(2) = \begin{bmatrix} -2.23 & 1.25 \\ -0.58 & 1.68 \end{bmatrix}, \quad B_{F22}(2) = \begin{bmatrix} 1.10 & 1.18 \\ -2.18 & 1.61 \end{bmatrix},$$

$$D(2) = \begin{bmatrix} 0.21 & 0.32 \\ 0.33 & 0.84 \end{bmatrix}, \quad E(2) = \begin{bmatrix} 2 & 3 \end{bmatrix},$$

$$R_{L11}(2) = 1.5I_2, \ R_{L12}(2) = 0.1I_2, \ R_{L21}(2) = 0.2I_2,$$

$$R_{L22}(2) = 1.5I_2, \ R_{F11}(2) = 3.0I_2, \ R_{F12}(2) = 0.3I_2,$$

$$R_{F21}(2) = 0.5I_2, \ R_{F22}(2) = 4.0I_2,$$

$$R_{LF11}(2) = \mathbf{diag}\begin{bmatrix} 1.5 & 1.3 \end{bmatrix}, \ R_{LF12}(2) = 0.2I_2,$$

$$R_{LF21}(2) = 0.6I_2, \; R_{LF22}(2) = \mathbf{diag}\begin{bmatrix}1.5 & 1.3\end{bmatrix},$$

$$R_{FL11}(2) = \mathbf{diag}\begin{bmatrix}1.2 & 1.4\end{bmatrix}, \; R_{FL12}(2) = 0.7I_2,$$

$$R_{FL21}(2) = 0.9I_2, \; R_{FL22}(2) = \mathbf{diag}\begin{bmatrix}2.5 & 1.1\end{bmatrix}.$$

The disturbance attenuation level is chosen as $\gamma = 5$. First, the leader's state-feedback Nash equilibrium strategies and the worst-case disturbance can be found through (88) using the following gain matrices:

$$K_{c1}^*(1) = \begin{bmatrix} K_{L11}^*(1) \\ K_{L12}^*(1) \\ K_{F11}^*(1) \\ K_{F21}^*(1) \end{bmatrix} = \begin{bmatrix} -1.2511\mathrm{e}\text{-}1 & 3.4787\mathrm{e}\text{-}2 \\ -1.3190\mathrm{e}\text{-}1 & -3.6915\mathrm{e}\text{-}2 \\ \hline -3.9477\mathrm{e}\text{-}1 & 1.0612\mathrm{e}\text{-}1 \\ -1.2425 & -1.5746\mathrm{e}\text{-}2 \\ \hline -6.7352\mathrm{e}\text{-}2 & -1.6364\mathrm{e}\text{-}2 \\ 1.3711\mathrm{e}\text{-}1 & -7.6100\mathrm{e}\text{-}2 \\ \hline -1.7551\mathrm{e}\text{-}1 & -6.8776\mathrm{e}\text{-}3 \\ 7.8614\mathrm{e}\text{-}1 & -2.9681\mathrm{e}\text{-}1 \end{bmatrix},$$

$$K_{c2}^*(1) = \begin{bmatrix} K_{L21}^*(1) \\ K_{L22}^*(1) \\ K_{F12}^*(1) \\ K_{F22}^*(1) \end{bmatrix} = \begin{bmatrix} -4.5891\mathrm{e}\text{-}1 & -1.4210 \\ -1.7034\mathrm{e}\text{-}1 & -1.3860\mathrm{e}\text{-}3 \\ \hline -4.1391\mathrm{e}\text{-}2 & -6.2093\mathrm{e}\text{-}2 \\ -1.8151\mathrm{e}\text{-}1 & -2.0851\mathrm{e}\text{-}2 \\ \hline -3.6319\mathrm{e}\text{-}1 & -3.0927\mathrm{e}\text{-}2 \\ -2.6017\mathrm{e}\text{-}1 & -1.0957\mathrm{e}\text{-}1 \\ \hline -1.3436\mathrm{e}\text{-}1 & -2.3530\mathrm{e}\text{-}1 \\ -1.0653\mathrm{e}\text{-}1 & -1.6604\mathrm{e}\text{-}1 \end{bmatrix},$$

$$K_\gamma^*(1) = \begin{bmatrix} 2.3592\mathrm{e}\text{-}2 & 5.1681\mathrm{e}\text{-}3 \\ 1.8501\mathrm{e}\text{-}2 & 3.5905\mathrm{e}\text{-}3 \end{bmatrix},$$

$$K_{c1}^*(2) = \begin{bmatrix} K_{L11}^*(2) \\ K_{L12}^*(2) \\ K_{F11}^*(2) \\ K_{F21}^*(2) \end{bmatrix} = \begin{bmatrix} 1.2847\mathrm{e}\text{-}2 & -7.5001\mathrm{e}\text{-}2 \\ 4.6662\mathrm{e}\text{-}3 & 5.8566\mathrm{e}\text{-}2 \\ \hline -3.9759\mathrm{e}\text{-}1 & -5.6783\mathrm{e}\text{-}1 \\ -5.2046 & 2.3993\mathrm{e}\text{-}1 \\ \hline -5.1654\mathrm{e}\text{-}2 & 2.3091\mathrm{e}\text{-}2 \\ -1.0087\mathrm{e}\text{-}1 & -1.0581\mathrm{e}\text{-}1 \\ \hline 6.9680\mathrm{e}\text{-}1 & 2.1147\mathrm{e}\text{-}1 \\ -4.6061\mathrm{e}\text{-}1 & -3.4329\mathrm{e}\text{-}1 \end{bmatrix},$$

$$K_{c2}^*(2) = \begin{bmatrix} K_{L21}^*(2) \\ K_{L22}^*(2) \\ K_{F12}^*(2) \\ K_{F22}^*(2) \end{bmatrix} = \begin{bmatrix} -4.3616\mathrm{e}\text{-}3 & -1.6256 \\ 2.8844\mathrm{e}\text{-}1 & 2.1600 \\ \hline 4.4060\mathrm{e}\text{-}2 & 3.1919\mathrm{e}\text{-}1 \\ -8.5707\mathrm{e}\text{-}2 & -5.2935\mathrm{e}\text{-}2 \\ \hline -9.3627\mathrm{e}\text{-}2 & -1.6558\mathrm{e}\text{-}1 \\ 2.4192\mathrm{e}\text{-}1 & -1.4402 \\ \hline -1.5050\mathrm{e}\text{-}2 & 2.0614\mathrm{e}\text{-}1 \\ -1.0369\mathrm{e}\text{-}1 & -2.1858\mathrm{e}\text{-}1 \end{bmatrix},$$

$$K_\gamma^*(2) = \begin{bmatrix} 1.1869\mathrm{e}\text{-}2 & 1.2445\mathrm{e}\text{-}2 \\ 2.2604\mathrm{e}\text{-}2 & 2.8804\mathrm{e}\text{-}2 \end{bmatrix}.$$

Second, the incentive strategies (91) announced by the leaders can be determined as follows:

$$\Theta_{F11}(1) = \begin{bmatrix} -9.8465\mathrm{e}\text{-}1 & 5.8400 \\ & -1.1403 & 2.7016 \end{bmatrix},$$

$$\Theta_{F12}(1) = \begin{bmatrix} 3.1313\mathrm{e}\text{-}2 & 1.1850\mathrm{e}\text{-}1 \\ -1.9308\mathrm{e}\text{-}1 & 6.7643\mathrm{e}\text{-}1 \end{bmatrix},$$

$$\Theta_{F21}(1) = \begin{bmatrix} -6.6324\mathrm{e}\text{-}1 & -2.2493 \\ 3.8222 & 4.5183 \end{bmatrix},$$

$$\Theta_{F22}(1) = \begin{bmatrix} -2.1088\mathrm{e}+01 & -1.2491\mathrm{e}+01 \\ 2.0869 & 1.1717 \end{bmatrix}.$$

$$\Theta_{F11}(2) = \begin{bmatrix} 1.1700\mathrm{e}+01 & 1.8406\mathrm{e}+01 \\ 1.5957\mathrm{e}+01 & 2.6060\mathrm{e}+01 \end{bmatrix},$$

$$\Theta_{F12}(2) = \begin{bmatrix} 5.1500\mathrm{e}\text{-}2 & -5.2684\mathrm{e}\text{-}1 \\ 2.3004\mathrm{e}\text{-}2 & -5.3342\mathrm{e}\text{-}1 \end{bmatrix},$$

$$\Theta_{F21}(2) = \begin{bmatrix} -7.7338\mathrm{e}\text{-}1 & 7.4665\mathrm{e}\text{-}1 \\ -5.1018\mathrm{e}\text{-}1 & 2.9019\mathrm{e}\text{-}1 \end{bmatrix},$$

$$\Theta_{F22}(2) = \begin{bmatrix} -1.5696 & 1.7622 \\ -3.9576\mathrm{e}\text{-}1 & -5.9039\mathrm{e}\text{-}1 \end{bmatrix}.$$

In fact, it is easy to verify that the proposed incentive strategy can induce the follower to choose the desired leader's strategy. On the other hand, it should be noted that there does not always exist a solution set of Theorems 3 and 4. In this case, the designers need to declare that no strategy exists.

Finally, the time histories are depicted from Fig. 7. As a result, one can find that the asymptotic stability can be achieved even if the mode changes.

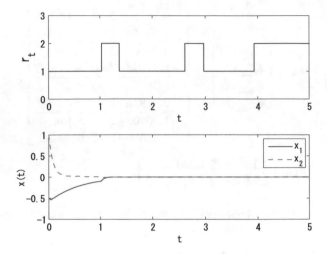

Fig. 7 Simulation result for state

4.8 *Incentive Possibility*

In this section, a novel concept for incentive possibility for the incentive is discussed for the special case. Incentive possibility is an important concept to guarantee an existence of the incentive such as controllability for the feedback control systems. Unless this condition holds, it is impossible to implement the incentive Stackelberg strategy to the practical plant.

The following special case of one leader and one follower is considered because it is easy to extend it to the general case. Let us consider the following MJLSS and the cost functions.

$$dx(t) = \left[A(r_t)x(t) + B(r_t)u(t) + D(r_t)v(t) \right] dt$$

$$+A_p(r_t)x(t)dw(t), \ x(0) = x^0, \tag{103a}$$

$$J_L(u, v) = J(u, v, Q_L), \tag{103b}$$

$$J_F(u, v) = J(u, v, Q_F), \tag{103c}$$

where

$$J(u, v, Q) = \frac{1}{2} \mathbb{E} \left[\int_0^\infty \left\{ x^T(t)Q(r_t)x(t) + u^T(t)R(r_t)u(t) \right. \right.$$

$$\left. \left. +v^T(t)R(r_t)v(t) \right\} dt \ \middle| \ r_0 = k \right].$$

It should be noted that all of the weight matrices for the controls are the same. In this case, the following relations are derived.

$$[D(k) + B(k)\Xi(k)]^T[X(k) - P(k)] = 0, \tag{104a}$$

$$P(k)A(k) + A^T(k)P(k) + A_p^T(k)P(k)A_p(k)$$

$$+ \sum_{m=1}^{s} \pi_{km} P(m) + P(k)\left[B(k) \ D(k) \right]$$

$$\times R^{-1}(k)\left[B(k) \ D(k) \right]^T P(k) + Q_L(k) = 0, \tag{104b}$$

$$X(k)A_C(k) + A_c^T(k)X(k) + A_p^T(k)X(k)A_p(k)$$

$$+ \sum_{m=1}^{s} \pi_{km} X(m) - [X(k)D_E(k) + S(k)][R_E(k)]^{-1}$$

$$\times[X(k)D_E(k) + S(k)]^T + Q_C(k) = 0, \tag{104c}$$

$$u(k) = \Lambda(k)x(k) + \Xi(k)v(k), \tag{104d}$$

where

$$\Lambda(k) := K(k) - \Xi(k)F(k),$$

$$F(k) := -[R_E(k)]^{-1}[X(k)D_E(k) + S_E(k)]^T,$$

$$F(k) := -R^{-1}(k)D^T(k)P(k),$$

$$K(k) := -R^{-1}(k)B^T(k)P(k),$$

$$A_C(k) := A(k) + B(k)\Lambda(k),$$

$$D_E(k) := D(k) + B(k)\Xi(k),$$

$$Q_C(k) := Q(k) + \Lambda^T(k)R(k)\Lambda(k),$$

$$S_E(k) := \Lambda^T(k)R(k)\Xi(k),$$

$$R_E(k) := R(k) + \Xi^T(k)R(k)\Xi(k).$$

First, if the dimension of $x(t)$ and $u(t)$ is $m = n$ and $B^{-1}(k)$ exists, the incentive $\Xi(k)$ can be solved as follows:

$$\Xi(k) = -B^{-1}(k)D(k). \tag{105}$$

In such a case, the MJLSS (103a) can be changed as follows:

$$dx(t) = A(r_t)x(t)dt + A_p(r_t)x(t)dw(t), \ x(0) = x^0. \tag{106}$$

It is obvious that the obtained MJLSS (106) is autonomous and uncontrollable. In particular, this is called the impossible incentive. Second, if $X(k) - P(k) = 0$ holds, then the following CCSAREs (107) can be derived from the CCSAREs in (104c).

$$P(k)A(k) + A^T(k)P(k) + A_p^T(k)P(k)A_p(k)$$

$$+ \sum_{m=1}^{s} \pi_{km} P(m) + P(k)\big[B(k) \ D(k) \big]$$

$$\times R^{-1}(k)\big[B(k) \ D(k) \big]^T P(k) + Q_F(k) = 0. \tag{107}$$

Therefore, if $Q_L(k) \neq Q_F(k)$, then $X(k) \neq P(k)$ holds. The fact seems to contradict this result. On the other hand, if $Q_L(k) = Q_F(k)$, all weights for the leader's and followers' cost functions in (103b) are the same. This is a weaker solution concept in team problems, the so-called person-by-person optimality [1] (p. 196). In this case, since $X(k) = P(k)$ holds, Eq. (104a) is always satisfied, the incentive is meaningless. Consequently, it should be pointed out that when the control designers consider the incentive Stackelberg game, they should pay special attention to the weight matrices of the cost function. It should be noted that the same problem will hold for the multiple leader and follower.

5 Static Output Feedback Case

In this section, the incentive Stackelberg game for a class of Markov jump linear stochastic systems (MJLSSs) with external disturbance is addressed. It should be noted that most of the results in this section are novel. In contrast to the previous sections, the static output feedback (SOF) incentive Stackelberg strategy with H_∞ constraint is studied for the first time.

5.1 Preliminary Results

First, the definitions of stochastic stabilizability and stochastic detectability, which are essential assumptions in the paper is introduced.

Definition 6 ([45]) Consider the following linear stochastically controlled system with Markovian jumps:

$$dx(t) = [A(r_t)x(t) + B(r_t)u(t)]dt + A_p(r_t)x(t)dw(t), \tag{108a}$$

$$y(t) = C(r_t)x(t), \tag{108b}$$

where $y(t) \in \mathbb{R}^p$ represents the output measurement vector.

First, system (108) or (A, B, A_p) is called stochastic stabilizable (in mean-square sense) by SOF, if there exists a feedback control $u(t) = F(r_t)y(t) = F(r_t)C(r_t)x(t)$ with $F(1), F(2), \ldots, F(s)$ being constant matrices, such that for any initial state $x(0) = x_0, r_0 = k$, the closed-loop system

$$
dx(t) = \left[A(r_t) + B(r_t)F(r_t)C(r_t)\right]x(t)dt
$$
$$
+A_p(r_t)x(t)dw(t), \quad x(0) = x^0, \tag{109}
$$

is AMSS, i.e.

$$
\lim_{t \to \infty} \mathbb{E}\left[\|x(t)\|^2 \mid r_0 = k\right] = 0. \tag{110}
$$

Second, under the condition that $B(r_t) \equiv 0$ that means autonomous systems, (A, A_p) is called stable, if Eq. (110) holds.

Definition 7 ([45]) The following state-measurement system:

$$
dx(t) = A(r_t)x(t)dt + A_p(r_t)x(t)dw(t), \tag{111a}
$$
$$
y(t) = C(r_t)x(t) \tag{111b}
$$

or $(A, A_p|C)$ is called stochastically detectable, if there exists a constant matrix X such that $(A + XC, A_p)$ is asymptotically mean-square stable.

In [45], it should be noted that necessary and sufficient conditions for stochastic stabilizability and stochastic detectability are provided in terms of solvability of some systems LMIs. Next, some useful lemmas are introduced.

Lemma 5 ([46, 47]) *If $(A, A_p|C)$ is stochastic detectable, then (A, A_p) is stable iff the stochastic algebraic Lyapunov equation (SALE) (112a) has a unique positive semi-definite solution P. Moreover, under the assumption that the Markov jump stochastic system (111a) is governed, (112b) holds.*

$$
P(k)A(k) + A^T(k)P(k) + A_p^T(k)P(k)A_p(k)
$$
$$
+ \sum_{\ell=1}^{s} \pi_{k\ell}P(\ell) + C^T(k)C(k) = 0, \tag{112a}
$$
$$
\mathbb{E}\left[\int_0^\infty x^T(t)C^T(r_t)C(r_t)x(t)dt \mid r_0 = k\right]
$$
$$
= \mathbb{E}\left[x^T(0)P(r_0)x(0) \mid r_0 = k\right] = \mathbb{E}[x^T(0)P(k)x(0)]. \tag{112b}
$$

The following lemma is an extension of the existing LQ control problem via the state feedback for infinite-horizon Markov jump linear stochastic systems. Furthermore, the proof can be done by using the above-mentioned Lemma 5.

Lemma 6 *Consider stochastic linear quadratic (LQ) control with jumps in the following form:*

$$\min_{u(t)} [J(u(t), x^0, k)], \quad \text{s.t.} \quad (108), \tag{113a}$$

$$u(t) = F(r_t)y(t) = F(r_t)C(r_t)x(t), \tag{113b}$$

where

$$J(u(t), x^0, k) = \mathbb{E}\left[\int_0^\infty \left[x^T(t)Q(r_t)x(t) + 2x^T(t)L(r_t)u(t) \right.\right.$$

$$\left.\left. + u^T(t)R(r_t)u(t) \right] dt \mid r_0 = k \right]. \tag{114}$$

Suppose that $x(0)$ is a zero mean random variable satisfying $\mathbb{E}[x(0)x^T(0)] = I_n$. Assume that there exist P and G that satisfy the following cross-coupled Lyapunov type equations (CCSALTEs) and that $C(k)G(k)C^T(k)$ is nonsingular.

$$P(k)[A(k) + B(k)F(k)C(k)] + [A(k) + B(k)F(k)C(k)]^T P(k)$$

$$+ A_p^T(k)P(k)A_p(k) + \sum_{\ell=1}^s \pi_{k\ell} P(\ell)$$

$$+ Q(k) + 2L(k)F(k)C(k) + C^T(k)F^T(k)R(k)F(k)C(k) = 0, \tag{115a}$$

$$G(k)[A(k) + B(k)F(k)C(k)]^T + [A(k) + B(k)F(k)C(k)]G(k)$$

$$+ A_p(k)G(k)A_p^T(k) + \sum_{\ell=1}^s \pi_{\ell k} G(\ell) + I_n = 0. \tag{115b}$$

Then, SOF control is given below.

$$u(t) = u^*(t) = F^*(r_t)y(t) = F^*(r_t)C(r_t)x(t), \tag{116}$$

where

$$F^*(k) = -R^{-1}(k)\left[B^T(k)P(k) + L^T(k) \right]G(k)C^T(k)$$

$$\times \left[C(k)G(k)C^T(k) \right]^{-1}.$$

Moreover,

$$J(u(t), x^0, k) \geq J(u^*(t), x^0, k) = \textbf{Trace}[P(k)]. \tag{117}$$

Consider the following stochastic linear system with Markovian jumps:

$$dx(t) = \left[A(r_t)x(t) + D(r_t)v(t)\right]dt + A_p(r_t)x(t)dw(t), \tag{118a}$$

$$z(t) = E(r_t)x(t), \tag{118b}$$

where $v(t) \in \mathbb{R}^{n_v}$ and $z(t) \in \mathbb{R}^{n_z}$ represent the external disturbance and controlled output, respectively.

The following result is already known as a bounded real lemma.

Lemma 7 ([45]) *For a given constant $\gamma > 0$, suppose there exists a symmetric non-negative definite solution $\mathbf{Z}$ to the following cross-coupled stochastic algebraic Riccati equations (CCSAREs)*

$$Z(k)A(k) + A^T(k)Z(k) + A_p^T(k)Z(k)A_p(k) + \sum_{\ell=1}^{s} \pi_{k\ell} Z(\ell)$$

$$+ \gamma^{-2} Z(k) D(k) D^T(k) Z(k) + E^T(k)E(k) = 0. \tag{119}$$

Then,

(i) *The stochastic linear system with Markov jumps* (118) *is asymptotically mean-square stable internally.*

(ii)

$$\|L\|_\infty^2 = \sup_{\substack{v \in L_F^2([0,\,\infty),\,\mathbb{R}^{n_v}), \\ v \neq 0,\, x^0 = 0}} \frac{\tilde{J}_1}{J_2} < \gamma, \tag{120}$$

where

$$v(t) = v^*(t) = \bar{F}_\gamma^*(r_t)x(t) = \gamma^{-2} D^T(r_t) Z(r_t) x(t),$$

$$\tilde{J}_1 := \sum_{i=1}^{s} \mathbb{E}\left[\int_0^\infty \|z(t)\|^2 dt \mid r_0 = k\right],$$

$$J_2 := \sum_{i=1}^{s} \mathbb{E}\left[\int_0^\infty \|v(t)\|^2 dt \mid r_0 = k\right].$$

It should be noted that $v^*(t)$ is called the worst-case disturbance.

In this paper, the SOF incentive Stackelberg strategy under H_∞ constraint for a class of stochastic MJLSS with multiple decision makers is solved.

5.2 Problem Formulation

Consider the following MJLSS with multiple leaders and followers:

$$dx(t) = \left[A(r_t)x(t) + \sum_{i=1}^{M} \sum_{j=1}^{N} \left[B_{Lij}(r_t)u_{Lij}(t) + B_{Fji}(r_t)u_{Fji}(t) \right] \right.$$

$$\left. + D(r_t)v(t) \right] dt + A_p(r_t)x(t)dw(t), \quad x(0) = x^0, \tag{121a}$$

$$z(t) = \mathbf{col} \left[E(r_t)x(t) \ u_{c1}(t) \ u_{c2}(t) \ \cdots \ u_{cM}(t) \right], \tag{121b}$$

$$y_{ci}(t) = C_{ci}(r_t)x(t), \tag{121c}$$

with

$$u_{ci}(t) = \mathbf{col} \left[u_{Li1}(t) \ \cdots \ u_{LiN}(t) \ u_{F1i}(t) \ \cdots \ u_{FNi}(t) \right],$$

$$y_{ci}(t) = \mathbf{col} \left[y_{Li1}(t) \ \cdots \ y_{LiN}(t) \ y_{F1i}(t) \ \cdots \ y_{FNi}(t) \right],$$

$$C_{ci}(r_t) = \mathbf{block \ diag} \left(C_{Li1}(r_t) \ \cdots \ C_{LiN}(r_t) \ C_{F1i}(r_t) \ \cdots \ C_{FNi}(r_t) \right),$$

where $u_{Lij}(t) \in \mathbb{R}^{m_{Lij}}$ represents the leader L_i's $i = 1, \ldots, M$ control input for the follower F_j. $u_{Fji}(t) \in \mathbb{R}^{m_{Fji}}$ represents the follower F_j's $j = 1, \ldots, N$ control input for the leader L_i in the sense of incentive Stackelberg strategy. $v(t) \in \mathbb{R}^{n_v}$ represents the external disturbance. $y_{Lij}(t) \in \mathbb{R}^{p_{Lij}}$ represents the leader L_i's output measurement vector. $y_{Fji}(t) \in \mathbb{R}^{p_{Fji}}$ represents the follower L_i's output measurement vector. The coefficients $\mathbf{A}, \mathbf{B}_{Lij}, \mathbf{B}_{Fji}, \mathbf{E}, \mathbf{A}_p, \mathbf{C}_{Lij}, \mathbf{C}_{Fji}$ are constant matrices of compatible dimensions, $i = 1, \ldots, M$, $j = 1, \ldots, N$.

Cost functionals of the leader L_i and the follower F_j are defined as follows:

$$J_{Li} \left(u_{c1}, \ldots, u_{cM}, v; x^0, k \right)$$

$$= \mathbb{E} \left[\int_0^\infty \left\{ x^T(t)Q_{Li}(r_t)x(t) + \sum_{j=1}^{N} \left[u_{Lij}^T(t)R_{Lij}(r_t)u_{Lij}(t) \right. \right. \right.$$

$$\left. \left. \left. + u_{Fji}^T(t)R_{LFji}(r_t)u_{Fji}(t) \right] \right\} dt \ \middle| \ r_0 = k \right], \quad i = 1, \ldots, M, \tag{122a}$$

$$J_{Fj} \left(\hat{u}_{F1}, \ldots, \hat{u}_{FN}, v; x^0, k \right)$$

$$= \mathbb{E} \left[\int_0^\infty \left\{ x^T(t)Q_{Fj}(r_t)x(t) + \sum_{i=1}^{M} \left[u_{Lij}^T(t)R_{FLij}(r_t)u_{Lij}(t) \right. \right. \right.$$

$$+ u_{Fji}^T(t) R_{Fji}(r_t) u_{Fji}(t) \Big] \Big\} dt \mid r_0 = k \Big], \quad j = 1, \dots, N, \qquad (122b)$$

where

$$\hat{u}_{Fj}(t) = \mathbf{col} \big[u_{Fj1}(t) \ \cdots \ u_{FjM}(t) \big],$$

$$Q_{Li}(k) = Q_{Li}^T(k) \geq 0, \ Q_{Fj}(k) = Q_{Fj}^T(k) \geq 0,$$

$$R_{Lij}(k) = R_{Lij}^T(k) > 0, \ R_{Fji} = R_{Fji}^T(k) > 0,$$

$$R_{LFji}(k) = R_{LFji}^T(k) \geq 0, \ R_{FLij}(k) = R_{FLij}^T(k) \geq 0,$$

$$k \in \mathcal{D}, \ i = 1, \dots, M, \ j = 1, \dots, N.$$

For an incentive Stackelberg game, leaders announce the following incentive strategy to the followers ahead of time:

$$u_{Lij}(t) = F_{Lij}(r_t) C(r_t) x(t) + \Xi_{ji}(r_t) \big[u_{Fji}(t) - F_{Lji}(r_t) C(r_t) x(t) \big]$$

$$= \Lambda_{ji}(r_t) x(t) + \Xi_{ji}(r_t) u_{Fji}(t), \qquad (123)$$

where $\Lambda_{ji}(k) = F_{Lij}^*(k) C(k) - Xi_{ji}(k) F_{Lji}^*(k) C(k)$.

The parameters $\Lambda_{ji}(k)$ and $\Xi_{ji}(k)$ are to be determined as associated with the Pareto optimal strategies $u_{Fji}(t)$ of the followers for $k \in \mathcal{D}, \ i = 1, \dots, M,$ $j = 1, \dots, N$. In this game, leaders will achieve a Pareto optimal strategy attenuating the external disturbance $v(t)$ with an H_∞ constraint. Infinite-horizon multiple leader-follower incentive Stackelberg games for MJLSSs with an H_∞ constraint can be formulated as follows:

For a given disturbance attenuation level $\gamma > 0$, find, if possible, the SOF controls

$$u_{Lij}^*(t) = F_{Lij}^*(r_t) C_{Lij}(r_t) x(t), \qquad (124a)$$

$$u_{Fji}^*(t) = F_{Fji}^*(r_t) C_{Fji}(r_t) x(t), \qquad (124b)$$

such that the following hold:

1. The MJLSS (121) attains the minimization of the centralized cost of (125a) with an H_∞ constraint condition (125b):

$$J_\tau(u_{c1}, \dots, u_{cM}, v^*; x^0, k) = \sum_{j=1}^{N} \tau_j J_{Li}(u_{c1}, \dots, u_{cM}, v^*; x^0, k), \qquad (125a)$$

$$0 \leq J_\gamma(u_{c1}^*, \dots, u_{cM}^*, v^*; x^0, k) \leq J_\gamma(u_{c1}^*, \dots, u_{cM}^*, v; x^0, k), \qquad (125b)$$

where $i = 1, \dots, M$ and

$$J_\tau(u_{c1}^*, \ldots, u_{cM}^*, v^*; x^0, k) = \min_{u_{c1}, \ldots, u_{cM}} J_\tau(u_{c1}, \ldots, u_{cM}, v^*; x^0, k),$$

$$\sum_{j=1}^{M} \tau_j = 1, \; 0 < \tau_j < 1, \; j = 1, \ldots, M,$$

$$J_\gamma(u_{c1}, \ldots, u_{cM}, v; x^0, k)$$

$$= \mathbb{E}\left[\int_0^\infty \left\{ \gamma^2 \|v(t)\|^2 - \|z(t)\|^2 \right\} dt \mid r_0 = k \right], \; v(t) \neq 0,$$

$$\|z(t)\|^2 = x^T(t) E^T(r_t) E(r_t) x(t) + \sum_{i=1}^{M} u_{ci}^T(t) u_{ci}(t).$$

On the other hand, consider the leader's incentive strategy (123) and the worst-case disturbance $v^*(t) \in \mathcal{L}_\mathcal{F}^2(\mathbb{R}_+, \mathbb{R}^{n_v})$. The follower's decision $u_{Fji}^*(t)) \in \mathcal{L}_\mathcal{F}^2(\mathbb{R}_+, \mathbb{R}^{n_{Fji}})$, $j = 1, \ldots, N$ can be selected as follows:

2. If the Pareto optimal strategy as cooperative strategy is chosen, the following objective function should be minimized.

$$\hat{J}_\rho(\hat{u}_{F1}, \ldots, \hat{u}_{FN}, v; x^0, k) = \sum_{j=1}^{N} \rho_j J_{Fj}(\hat{u}_{F1}, \ldots, \hat{u}_{FN}, v; x^0, k), \qquad (126)$$

where $\sum_{j=1}^{N} \rho_j = 1, 0 < \rho_j < 1, j = 1, \ldots, N$.

5.3 Main Results

In this section, the Pareto optimal strategies of the leaders and the Pareto strategies of the followers are derived.

5.3.1 Leader's Pareto Optimal Strategy

It is assumed that leaders are cooperative within their group, and they will find the Pareto optimal strategy. Therefore, the Pareto optimal solutions $(u_{c1}^*(t), \ldots, u_{cM}^*(t), v^*(t))$ of the leaders are investigated in terms of how they attenuate the disturbance under an H_∞ constraint. For this purpose, let us configure the MJLSS (127) as the following centralized system:

$$dx(t) = \left[A(r_t)x(t) + \sum_{i=1}^{M} B_{ci}(r_t)u_{ci}(t) + D(r_t)v(t) \right] dt$$

$$+ A_p(r_t)x(t)dw(t), \quad x(0) = x^0, \tag{127a}$$

$$z(t) = \mathbf{col}\left[E(r_t)x(t)\ u_{c1}(t)\ u_{c2}(t)\ \cdots\ u_{cM}(t) \right], \tag{127b}$$

$$u_{ci}(t) = F_{ci}(r_t)y_{ci}(t) = F_{ci}(r_t)C_{ci}(r_t)x(t), \tag{127c}$$

where $i = 1, \ldots, M$ and $k \in \mathcal{D}$.

$$B_{ci}(k) = \left[B_{Li1}(k) \ \cdots \ B_{LiN}(k)\ B_{F1i}(k) \ \cdots \ B_{FNi}(k) \right],$$

$$F_{ci}^*(k) = \mathbf{block\ diag}\left(F_{Li1}^*(k) \ \cdots \ F_{LiN}^*(k)\ F_{F1i}^*(k) \ \cdots \ F_{FNi}^*(k) \right).$$

Furthermore, the cost functional (122a) can be changed as follows:

$$J_{Li}(u_{c1}, \ldots, u_{cM}; x^0, k)$$

$$= \mathbb{E}\left[\int_0^\infty \left\{ x^T(t)Q_{Li}(r_t)x(t) + u_{ci}^T(t)R_{ci}(r_t)u_{ci}(t) \right\}dt \Big| r_0 = k \right], \tag{128}$$

where

$$R_{ci}(k) = \mathbf{block\ diag}\left(R_{Li1}(k) \ \cdots \ R_{LiN}(k)\ R_{LF1i}(k) \ \cdots \ R_{LFNi}(k) \right).$$

To obtain the Pareto optimal strategy of the ith leader under the H_∞ constraint, the following results are provided by using Lemma 6.

Corollary 2 *For a given disturbance attenuation level* $\gamma > 0$, *suppose that there exist* $\boldsymbol{P}_c$, $\boldsymbol{G}_c$, *and* $\boldsymbol{Z}$ *that satisfy the following cross-coupled Lyapunov type equations (CCSALTEs) and that* $C_{ci}(k)G_c(k)C_{ci}^T(k)$ *is nonsingular.*

$$P_c(k)A_{\gamma c}(k) + A_{\gamma c}^T(k)P_c(k) + A_p^T(k)P_c(k)A_p(k) + \sum_{\ell=1}^{s} \pi_{k\ell} P_c(\ell)$$

$$+ \sum_{j=1}^{M} \tau_j\left[Q_{Lj}(k) + C_{cj}^T(k)F_{cj}^T(k)R_{cj}(k)F_{cj}(k)C_{cj}(k) \right] = 0, \tag{129a}$$

$$G_c(k)A_{\gamma c}^T(k) + A_{\gamma c}(k)G_c(k) + A_p(k)G_c(k)A_p^T(k)$$

$$+ \sum_{\ell=1}^{s} \pi_{\ell k}G_c(\ell) + I_n = 0, \tag{129b}$$

$$Z(k)A_c(k) + A_c^T(k)Z(k) + A_p^T(k)Z(k)A_p(k) + \sum_{\ell=1}^{s} \pi_{k\ell}Z(\ell)$$

$$+\gamma^{-2}Z(k)D(k)D^T(k)Z(k) + E_c^T(k)E_c(k) = 0, \tag{129c}$$

where

$$A_{\gamma c}(k) = A(k) + \sum_{j=1}^{M} B_{ci}(k) F_{cj}(k) C_{cj}(k) + \gamma^{-2} D(k) D^T(k) Z(k),$$

$$A_c(k) = A(k) + \sum_{j=1}^{M} B_{cj}(k) F_{cj}(k) C_{cj}(k),$$

$$F_{Lij}(k) = -[\tau_i R_{Lij}]^{-1}(k) B_{Lij}^T(k) P_c(k) G_c(k) C_{Lij}^T(k)$$

$$\times \left[C_{Lij}(k) G_c(k) C_{Lij}^T(k) \right]^{-1},$$

$$F_{Fji}(k) = -[\tau_i R_{LFji}]^{-1}(k) B_{Fji}^T(k) P_c(k) G_c(k) C_{Fji}^T(k)$$

$$\times \left[C_{Fji}(k) G_c(k) C_{Fji}^T(k) \right]^{-1},$$

$$E_c(k) = \begin{bmatrix} E(k) \\ F_{c1}(k) C_{c1}(k) \\ \vdots \\ F_{cM}(k) C_{cM}(k) \end{bmatrix}, \ i = 1, \ldots, M \ \text{and} \ k \in \mathcal{D}.$$

Then the H_∞ constraint SOF control problem has the following solutions:

$$u_{ci}(t) = u_{ci}^*(t) = F_{ci}(r_t) y(t) = F_{ci}^*(r_t) y(t), \tag{130a}$$

$$v(t) = v^*(t) = F_\gamma(r_t) x(t) = F_\gamma^*(r_t) x(t) = \gamma^{-2} D^T(r_t) Z(r_t) x(t). \tag{130b}$$

5.3.2 Follower's Pareto Optimal Strategy

In this subsection, each follower's strategy associated with the incentive strategy (123) and the incentive parameters $\Xi_{ij}(r_t)$ is thereby established. Substituting (123) together with Nash strategy set (130) into MJLSS (121a) and the cost function (122b), the following optimization problem can be obtained:

$$\min_{\mathbb{F}_F} J_\rho(u_{F1}, \ldots, u_{FN}, v; x^0, k)$$

$$= \mathbb{E}\left[\int_0^\infty \left[x^T(t) Q_{F\rho}(r_t) x(t) + 2x^T(t) L_{F\rho}(r_t) u_F(t) \right.\right.$$

$$\left.\left. + u_F^T(t) R_{F\rho}(r_t) u_F(t) \right] dt \, \middle| r_0 = k \right], \tag{131a}$$

s.t.

$$dx(t) = \left[A_{\mathrm{F}c}(r_t)x(t) + B_{\mathrm{F}c}(r_t)u_{\mathrm{F}}(t) \right] dt$$

$$+ A_p(r_t)x(t)dw(t), \ x(0) = x^0, \tag{131b}$$

$$u_{\mathrm{F}}(t) = \begin{bmatrix} u_{\mathrm{F}1} \\ \vdots \\ u_{\mathrm{F}N} \end{bmatrix} = F_{\mathrm{F}c}(r_t)y_{\mathrm{F}c}(t) = F_{\mathrm{F}c}(r_t)C_{\mathrm{F}c}(r_t)x(t), \tag{131c}$$

$$y_{\mathrm{F}c}(t) = \begin{bmatrix} y_{\mathrm{F}1}(t) \\ \vdots \\ y_{\mathrm{F}N}(t) \end{bmatrix}, \ y_{\mathrm{F}j}(t) = \begin{bmatrix} y_{\mathrm{F}j1}(t) \\ \vdots \\ y_{\mathrm{F}jM}(t) \end{bmatrix},$$

$$y_{\mathrm{F}j1}(t) = C_{\mathrm{F}j1}x(t), \tag{131d}$$

where

$$Q_{\mathrm{F}\rho}(k) := \sum_{j=1}^{N} \rho_j \left[Q_{\mathrm{F}j}(k) + \sum_{i=1}^{M} \Lambda_{ji}^{T}(k) R_{\mathrm{FL}ij}(k) \Lambda_{ji}(k) \right],$$

$$L_{\mathrm{F}\rho}(k) := \left[\rho_1 L_{\mathrm{F}1}(k) \cdots \rho_N L_{\mathrm{F}N}(k) \right],$$

$$L_{\mathrm{F}j}(k) := \left[\Lambda_{j1}^{T}(k) R_{\mathrm{FL}1j}(k) \Xi_{j1}(k) \right.$$

$$\left. \cdots \Lambda_{jM}^{T}(k) R_{\mathrm{FL}Mj}(k) \Xi_{jM}(k) \right],$$

$$R_{\mathrm{F}\rho}(k) := \textbf{block diag} \left(\rho_1 R_{\mathrm{F}1}(k) \cdots \rho_N R_{\mathrm{F}N}(k) \right),$$

$$R_{\mathrm{F}j}(k) := \textbf{block diag} \left(R_{\mathrm{F}j1}(k) + \Xi_{j1}^{T}(k) R_{\mathrm{FL}1j}(k) \Xi_{j1}(k) \right.$$

$$\left. \cdots R_{\mathrm{F}jM}(k) + \Xi_{jM}^{T}(k) R_{\mathrm{FL}Mj}(k) \Xi_{jM}(k) \right),$$

$$A_{\mathrm{F}c}(k) := A(k) + \sum_{i=1}^{M} \sum_{j=1}^{N} B_{\mathrm{L}ij}(k) \Lambda_{ji}(k) + \gamma^{-2} D D^{T}(k) Z(k),$$

$$B_{\mathrm{F}c}(k) := \left[B_{\mathrm{F}1}(k) \cdots B_{\mathrm{F}N}(k) \right],$$

$$B_{\mathrm{F}j}(k) := \left[B_{\mathrm{F}cj1}(k) \cdots B_{\mathrm{F}cjM}(k) \right],$$

$$B_{\mathrm{F}cij}(k) := B_{\mathrm{F}ij}(k) + B_{\mathrm{L}ji}(k) \Xi_{ij}(k).$$

Using Lemma 6 to the above-mentioned optimization problem, the Pareto optimal strategy set can be obtained as follows:

$$P_{\mathrm{F}}(k)\big[A_{\mathrm{F}c}(k) + B_{\mathrm{F}c}(k)F_{\mathrm{F}c}(k)C_{\mathrm{F}c}(k)\big]$$

$$+\big[A_{\mathrm{F}c}(k) + B_{\mathrm{F}c}(k)F_{\mathrm{F}c}(k)C_{\mathrm{F}c}(k)\big]^{T}P_{\mathrm{F}}(k)$$

$$+A_{p}^{T}(k)P_{\mathrm{F}}(k)A_{p}(k) + \sum_{\ell=1}^{s}\pi_{k\ell}P_{\mathrm{F}}(\ell)$$

$$+L_{\mathrm{F}\rho}(k)F_{\mathrm{F}c}(k)C_{\mathrm{F}c}(k) + C_{\mathrm{F}c}^{T}(k)F_{\mathrm{F}c}^{T}(k)L_{\mathrm{F}\rho}^{T}(k) + Q_{\mathrm{F}\rho}(k)$$

$$+C_{\mathrm{F}c}^{T}(k)F_{\mathrm{F}c}^{T}(k)R_{\mathrm{F}\rho}(k)F_{\mathrm{F}c}(k)C_{\mathrm{F}c}(k) = 0, \tag{132a}$$

$$G_{\mathrm{F}}(k)\big[A_{\mathrm{F}c}(k) + B_{\mathrm{F}c}(k)F_{\mathrm{F}c}(k)C_{\mathrm{F}c}(k)\big]^{T}$$

$$+\big[A_{\mathrm{F}c}(k) + B_{\mathrm{F}c}(k)F_{\mathrm{F}c}(k)C_{\mathrm{F}c}(k)\big]G_{\mathrm{F}}(k)$$

$$+A_{p}G_{\mathrm{F}}(k)A_{p}^{T}(k) + \sum_{\ell=1}^{s}\pi_{\ell k}G_{\mathrm{F}}(\ell) + I_{n} = 0. \tag{132b}$$

Then, SOF strategy based on the incentive (123) is given below.

$$u_{\mathrm{F}}(t) = u_{\mathrm{F}}^{\dagger}(t) = F_{\mathrm{F}c}^{\dagger}(r_{t})y_{\mathrm{F}c}(t) = F_{\mathrm{F}c}^{\dagger}(r_{t})C_{\mathrm{F}c}(r_{t})x(t), \tag{133}$$

where

$$F_{\mathrm{F}c}^{\dagger}(k) = \mathbf{block\ diag}\left(F_{\mathrm{F}1}^{\dagger}(k) \cdots F_{\mathrm{F}N}^{\dagger}(k)\right),$$

$$F_{\mathrm{F}j}^{\dagger}(k) = \mathbf{block\ diag}\left(F_{\mathrm{F}j1}^{\dagger}(k) \cdots F_{\mathrm{F}jM}^{\dagger}(k)\right),$$

$$F_{\mathrm{F}ji}^{\dagger}(k) = -\big[\rho_{j}\big(R_{\mathrm{F}ji}(k) + \Xi_{ji}^{T}(k)R_{\mathrm{F}Lij}(k)\Xi_{ji}(k)\big)\big]^{-1}$$

$$\times\big[\big(B_{\mathrm{F}ji}(k) + B_{\mathrm{L}ij}(k)\Xi_{ji}^{T}(k)\big)^{T}P_{\mathrm{F}}(k)$$

$$+\big(\rho_{j}\Lambda_{ji}^{T}(k)R_{\mathrm{F}Lij}(k)\Xi_{ji}^{T}(k)\big)^{T}\big]G_{\mathrm{F}}(k)C_{\mathrm{F}ji}^{T}(k)$$

$$\times\big[C_{\mathrm{F}ji}G_{\mathrm{F}}(k)C_{\mathrm{F}ji}^{T}(k)\big]^{-1}.$$

Moreover, $\Xi_{ij}(k)$, $i = 1, \ldots, M$, $j = 1, \ldots, N$ satisfy the following linear algebraic matrix equations (LAMEs):

$$\Xi_{ij}^{T}(k)\big(B_{\mathrm{L}ji}^{T}(k)P_{\mathrm{F}}(k) + R_{\mathrm{F}Lji}(k)F_{\mathrm{L}ji}^{*}(k)C_{\mathrm{L}ji}(k)\big)$$

$$+ R_{\mathrm{F}ij}(k)F_{\mathrm{F}ij}^{*}(k)C_{\mathrm{F}ij}(k) + B_{\mathrm{F}ij}^{T}(k)P_{\mathrm{F}}(k) = 0. \tag{134}$$

It should be noted that LAMEs (134) can be established by using the relation $F_{\mathrm{F}ji}^{*}(k) = F_{\mathrm{F}ji}^{\dagger}(k)$.

6 Incentive Stackelberg Strategy for Stochastic LPV Systems

In this section, the incentive Stackelberg-Nash strategy for stochastic LPV systems with disturbance attenuation is discussed under multiple decision makers. As compared to previous results, the incentive strategy for a class of stochastic LPV systems in [36, 38] was investigated here for the first time.

6.1 Preliminary Results

Consider the following stochastic LPV system:

$$dx(t) = \big[A(\theta(t))x(t) + Bu(t) + Dv(t)\big]dt$$
$$+\big[A_p x(t) + D_p v(t)\big]dw(t), \quad x(0) = x^0, \tag{135a}$$

$$z(t) = \begin{bmatrix} E(\theta(t))x(t) \\ Gu(t) \end{bmatrix}, \quad G^T G = I_n, \tag{135b}$$

where $x(t) \in \mathbb{R}^n$ denotes the state vector. $u(t) \in \mathbb{R}^m$ denotes the control input. $v(t) \in \mathbb{R}^{n_v}$ denotes the external disturbance. $z(t) \in \mathbb{R}^{n_z + m_z}$ ($E(\theta(t))x(t) \in \mathbb{R}^{n_z}$, $Gu(t) \in \mathbb{R}^{m_z}$) denotes the controlled output. $w(t) \in \mathbb{R}$ denotes a one-dimensional standard Wiener process defined in the filtered probability space [6, 49, 50]. $\theta(t) \in \mathbb{R}^r$ denotes the time-varying parameters. r is the number of time-varying parameters. Without loss of generality, it is assumed that the stochastic system (135) has a unique strong solution $x(t) = x(t, \tilde{u}, \tilde{v}, x(0))$ for any $u(t) = \tilde{u}(x(t))$ and $v(t) = \tilde{v}(x(t))$. The coefficient matrices $A(\theta(t))$ and $A_p(\theta(t))$ are parameter-dependent matrices that can be expressed as

$$A(\theta(t)) = \sum_{k=1}^{M} \alpha_k(t) A_k, \quad E(\theta(t)) = \sum_{k=1}^{M} \alpha_k(t) E_k, \tag{136}$$

where $\alpha_k(t) \geq 0$, $\sum_{k=1}^{M} \alpha_k(t) = 1$, $M = 2^r$ and r is the number of time-varying parameters [51].

It should be noted that the above-mentioned descriptions are used to simplify the context. Furthermore, for notational convenience, instead of $\theta(t)$, θ will be used, with similar abbreviations used subsequently.

The following definition of stochastic stability will be required.

Definition 8 ([51]) A stochastic LPV system governed by Itô's differential equation (135) is mean-square stable if the trajectories satisfy

$$\lim_{t \to \infty} \mathbb{E}[\|x(t)\|^2] = 0$$

for any initial condition.

The H_∞ norm, an essential assumption here, is introduced in [51].

Definition 9 ([51]) The H_∞ norm of the stochastic LPV system (135) with mean-square stable is given by

$$\|L\|_\infty^2 = \sup_{\substack{v \in L_F^2([0,\,\infty),\,\mathbb{R}^{n_v}), \\ v \neq 0,\, x^0 = 0}} \frac{J_z}{J_v}, \tag{137}$$

where

$$J_z := \mathbb{E}\left[\int_0^\infty \|z(t)\|^2 dt\right], \quad J_v := \mathbb{E}\left[\int_0^\infty \|v(t)\|^2 dt\right].$$

Lemma 8 ([36, 38]) *Let us consider an autonomous system* (135) *with* $u(t) \equiv 0$. *For a given attenuation performance level* $\gamma > 0$, *if there exists a matrix* $Z = Z^T > 0$ *satisfying the following linear matrix inequalities* (LMIs) (138), *the stochastic LPV system* (135) *is mean-square stable with* $\|L\|_\infty < \gamma$ *below* $x^0 = 0$.

$$\begin{bmatrix} \Psi_k(Z) & E_k^T & L_\gamma^T \\ E_k & -I_{n_z} & 0 \\ L_\gamma & 0 & -R_\gamma \end{bmatrix} < 0, \tag{138}$$

where $k = 1, \ldots, M$, $\Psi_k(Z) := ZA_k + A_k^T Z + A_p^T ZA_p$, $R_\gamma := \gamma^2 I_{n_v} - D_p^T ZD_p > 0$, $L_\gamma := D^T Z + D_p^T ZA_p$.
Moreover, the worst-case disturbance is given by

$$v^*(t) := R_\gamma^{-1} L_\gamma x(t). \tag{139}$$

On the other hand, the standard linear quadratic control (LQC) problem for a stochastic LPV system with $v(t) \equiv 0$ or $D \equiv 0$ in (135a) is given [32, 48].

Definition 10 ([32, 48]) Let us consider the stochastic LPV system with $v(t) \equiv 0$ in (135a). The following cost performance is defined by

$$J(u,\, x^0) = \mathbb{E}\left[\int_0^\infty \left[x^T(t)Qx(t) + u^T(t)Ru(t)\right]dt\right], \tag{140}$$

where $Q = Q^T > 0$, $R = R^T > 0$.
In this situation, the LQC problem is to find a fixed state feedback control

$$u(t) = Kx(t) \tag{141}$$

such that the quadratic cost functional (140) is minimized.

The following result can be obtained using a technique similar to that in [48].

Lemma 9 ([36, 38]) *If there exist a matrix $\bar{X} > 0$ and $\bar{Y}$ satisfying the LMIs* (142)

$$
\begin{bmatrix}
\bar{\Xi}_k(\bar{X}, \bar{Y}) & \bar{X} & \bar{Y}^T & \bar{X}A_p^T \\
\bar{X} & -Q^{-1} & 0 & 0 \\
\bar{Y} & 0 & -R^{-1} & 0 \\
A_p\bar{X} & 0 & 0 & -\bar{X}
\end{bmatrix} \le 0, \; k = 1, \dots, M,
\tag{142}
$$

where

$$
\bar{\Xi}_k(\bar{X}, \bar{Y}) := A_k\bar{X} + \bar{X}A_k^T + B\bar{Y} + \bar{Y}^T B^T,
$$

$$
K := \bar{Y}\bar{X}^{-1},
$$

then

$$
J(u, \, x^0) \le \mathbb{E}\big[x^T(0)\bar{X}^{-1}x(0)\big].
\tag{143}
$$

It should be noted that the result obtained corresponding to Lemmas 8 and 9 is a sufficient condition.

6.2   Problem Formulation

Consider the stochastic LPV system governed by Itô's differential equation with multiple decision makers defined by

$$
dx(t) = \left[A(\theta(t))x(t) + \sum_{j=1}^{N} \big[B_{0j}u_{0j}(t) + B_j u_j(t)\big] \right.
$$

$$
\left. + Dv(t) \right]dt + \big[A_p x(t) + D_p v(t)\big]dw(t),
\tag{144a}
$$

$$
z(t) = \begin{bmatrix} E(\theta(t))x(t) \\ \tilde{G}\tilde{u}(t) \end{bmatrix},
\tag{144b}
$$

where

$$
\tilde{G}\tilde{u}(t) := \begin{bmatrix} G_0 u_0(t) \\ G_1 u_1(t) \\ \vdots \\ G_N u_N(t) \end{bmatrix},
$$

$$
u_0(t) = \mathbf{col}\big[u_{01}(t) \cdots u_{0N}(t)\big],
$$

$$
u_{0i}(t) = K_{0i}x(t),
$$

$$u_i(t) = K_i x(t),$$

$$G_0 := \textbf{block diag} \left(G_{01} \cdots G_{0N} \right),$$

$$G_{0i} u_{0i}(t) \in \mathbb{R}^{m_{z0i}}, \ G_i u_i(t) \in \mathbb{R}^{m_{zi}}.$$

$u_0(t) \in \mathbb{R}^{m_0}$, $m_0 = \sum_{j=1}^{N} m_{0j}$ with $u_{0j}(t) \in \mathbb{R}^{m_{0j}}$ denotes the leader's control input. $u_i(t) \in \mathbb{R}^{m_i}$, $i = 1, \ldots, N$ denotes the ith follower's control input. In the following, we use P_0 to represent the leader and P_i, $i = 1, \ldots, N$ to represent the ith follower. Other variables are defined by stochastic equation (135). It should be noted that G_i does not depend on a time-varying parameter because the controlled output can be chosen by the controller designer. Hence, without loss of generality, assume that $G_i^T G_i = I_{m_{zi}}$, $i = 0, 1, \ldots, N$, $m_{z0} = \sum_{j=1}^{N} m_{z0j}$, $G_i \in \mathbb{R}^{g_i \times m_i}$. Furthermore, without loss of generality, to remove this dependence on $x(0)$, suppose that $x(0)$ is a zero mean random variable satisfying $\mathbb{E}[x(0)x^T(0)] = I_n$.

The cost performances are defined by

$$J_v(u_0, u_1, \ldots, u_N, \ v, \ x^0) = \mathbb{E}\left[\int_0^\infty \left[\gamma^2 \|v(t)\|^2 - \|z(t)\|^2 \right] dt \right], \tag{145a}$$

$$J_0(u_0, u_1, \ldots, u_N, \ v, \ x^0)$$

$$= \mathbb{E}\left[\int_0^\infty \left\{ x^T(t) Q_0 x(t) + \sum_{i=1}^{N} \left[u_{0i}^T(t) R_{00i} u_{0i}(t) \right. \right. \right.$$

$$\left. \left. \left. + u_i^T(t) R_{0i} u_i(t) \right] \right\} dt \right], \tag{145b}$$

$$J_i(u_1, \ldots, u_N, \ v, \ x^0)$$

$$= \mathbb{E}\left[\int_0^\infty \left\{ x^T(t) Q_i x(t) + u_i^T(t) R_{ii} u_i(t) \right\} dt \right], \tag{145c}$$

where $i = 1, \ldots, N$, $Q_0 := Q_0^T$, $Q_i = Q_i^T > 0$, $R_{00i} = R_{00i}^T > 0$, $R_{0i} = R_{0i}^T \geq 0$, $R_{ii} = R_{ii}^T > 0$.

At first glance, there seems to be no input from other players. However, since the state after implementing the feedback strategy depends on the control inputs of all the players, the left-hand side of the cost function can be represented as Eq. (145c).

The infinite-horizon incentive Stackelberg-Nash strategy with the H_∞ constraint for the stochastic LPV system (144) is defined as follows:

For any given $\gamma > 0$, find a fixed state feedback strategy set $u_i(t) = u_i^\dagger(t) = K_i^\dagger x(t) \in L_F^2([0, \infty), \mathbb{R}^{m_i})$, $i = 1, \ldots, N$ such that

1. *the leader announces a strategy that minimizes the cost function (145b) ahead of time to the followers with the following feedback pattern.*

$$u_{0i}(t) = u_{0i}(x, u_i) = \eta_{0i} x(t) + \eta_{ii} u_i(t) \tag{146}$$

for $i = 1, \ldots, N$, where $\eta_{0i} \in \mathbb{R}^{m_{0i} \times n}$ and $\eta_{ii} \in \mathbb{R}^{m_{0i} \times m_i}$ are strategy parameter matrices.

2. *For the closed-loop stochastic LPV system, H_∞ norm conditions hold such that $J_v(u_1^*, \ldots, u_N^*, v, x^0) \geq 0$ and $u_i(t) = u_i^*(t) = K_i^* x(t)$.*

3. *When the worst-case disturbance $v^*(t)$ is implemented in (144), $u_i(t) = u_i^\dagger(t)$, $i = 1, \ldots, N$ satisfies the following Nash equilibrium condition (147):*

$$J_i(u_1^*, \ldots, u_N^*, v^*, x^0)$$

$$\leq J_i(u_1^*, \ldots, u_{i-1}^*, u_i, u_{i+1}^*, \ldots, u_N^*, v^*, x^0). \tag{147}$$

In other words, first, solve the optimization problem (148a). Second, find $v(t) = v^*(t)$ such that inequality (148b) holds when the team-optimal strategy set $u_i(t) = u_i^*(t)$ is applied. Finally, find $\eta_{ii} = \eta_{ii}^\dagger$ and $u_i(t) = u_i^\dagger(t)$, where $u_{0i}(t) = u_{0i}^*(t) + \eta_{ii}(u_i(t) - u_i^*(t))$ such that the solution set of the optimization problem (148c) is $u_i^\dagger(t) = u_i^*(t)$.

$$J_0(u_0^*, u_1^*, \ldots, u_N^*, v^*, x^0)$$

$$= \min_{u_0, u_1, \ldots, u_N} J_0(u_0, u_1, \ldots, u_N, v^*, x^0), \tag{148a}$$

$$0 \leq J_v(u_0^*, u_1^*, \ldots, u_N^*, v^*, x^0) \leq J_v(u_0^*, u_1^*, \ldots, u_N^*, v, x^0), \tag{148b}$$

$$J_i(u_1^\dagger, \ldots, u_N^\dagger, v^*, x^0)$$

$$\leq J_i(u_1^\dagger, \ldots, u_{i-1}^\dagger, u_i, u_{i+1}^\dagger, \ldots, u_N^\dagger, v^*, x^0). \tag{148c}$$

The reasons for introducing a fixed gain while introducing the LPV system is based on the following two concerns. First, when the incentive has a variable structure, the optimization problem to be solved is a bilinear matrix inequality (BMI), which is very difficult to solve, or at worst, the solution might not exist. Second, for mounting on-board computers, such as smart meters, the next generation of electric meters, there are some cases in which only a fixed gain can be implemented because the control program must be downsized owing to the limited memory area and the performance limitations of the central processing unit.

In the next section, we derive the solution of the above-mentioned problem, the H_∞-constraint Nash strategy.

6.3 Main Results

First, the team-optimal solution set with the H_∞ constraint is derived. By centralizing the control inputs in the stochastic LPV system (144), the following centralized stochastic systems can be obtained:

$$dx(t) = \big[A(\theta(t))x(t) + B_c u_c(t) + Dv(t)\big]dt$$
$$+\big[A_p x(t) + D_p v(t)\big]dw(t), \tag{149a}$$

$$z(t) = \begin{bmatrix} E(\theta(t))x(t) \\ G_c u_c(t) \end{bmatrix}, \tag{149b}$$

$$u_c(t) = \mathbf{col}\big[\,u_0(t)\ u_1(t)\ \cdots\ u_N(t)\,\big], \tag{149c}$$

where

$$B_c := \big[\,B_0\ B_1\ \cdots\ B_N\,\big],$$
$$B_0 := \big[\,B_{01}\ \cdots\ B_{0N}\,\big],$$
$$G_c := \mathbf{block\ diag}\,\big(\,G_0\ G_1\ \cdots\ G_N\,\big).$$

Furthermore, the cost functional (145b) can be changed as:

$$J_0(u_0, u_1, \ldots, u_N,\ v,\ x^0)$$
$$= \mathbb{E}\bigg[\int_0^\infty \big\{x^T(t)Q_0 x(t) + u_c^T(t)R_c u_c(t)\big\}dt\bigg], \tag{150}$$

where

$$R_c := \mathbf{block\ diag}\,\big(\,R_{00}\ R_{01}\ \cdots\ R_{0N}\,\big),$$
$$R_{00} := \mathbf{block\ diag}\,\big(\,R_{001}\ \cdots\ R_{00N}\,\big).$$

Using Lemmas 8 and 9, the following conditions can be obtained.

Theorem 5 ([36]) *Let us consider the stochastic* LPV *system* (144) *with multiple decision makers* $u_i(t)$ *and the deterministic disturbance* $v(t)$. *For a given attenuation performance level* $\gamma > 0$, *assume that there exists a solution set for the real symmetric matrices* $X > 0$, Y_k, *and* $W > 0$ *such that the following* CCMIs *are satisfied:*

$$\begin{bmatrix} \boldsymbol{\Xi}_k(X,Y) & X & Y^T & X\bar{A}_p^T \\ X & -Q_0^{-1} & 0 & 0 \\ Y & 0 & -R_c^{-1} & 0 \\ \bar{A}_p X & 0 & 0 & -X \end{bmatrix} \leq 0, \tag{151a}$$

$$\begin{bmatrix} \boldsymbol{\Theta}_k(W) & E_{Kk}^T & L_{\gamma W}^T \\ E_{Kk} & -I_{n_z} & 0 \\ L_{\gamma W} & 0 & -R_{\gamma W} \end{bmatrix} < 0, \tag{151b}$$

where $k = 1, \ldots, M,$

$$\Xi_k(X, Y) := A_{F_\gamma k} X + X A_{F_\gamma k}^T + B_c Y + Y^T B_c^T,$$

$$\Theta_k(W) := W A_{Kk} + A_{Kk}^T W + A_p^T W A_p,$$

$$A_{F_\gamma k} := A_k + D F_\gamma^*, \quad \bar{A}_p = A_p + D_p F_\gamma^*,$$

$$A_{Kk} := A_k + \sum_{j=1}^{N} [B_{0j} K_{0j}^* + B_j K_j^*],$$

$$\bar{z}_n := n_z + \sum_{j=1}^{N} (m_{z0j} + m_{zj}),$$

$$E_{Kk} := \begin{bmatrix} E_k^T \\ G_0 K_0^* \\ G_1 K_1^* \\ \vdots \\ G_N K_N^* \end{bmatrix},$$

$$K^* := Y X^{-1} = \begin{bmatrix} K_0^* \\ K_1^* \\ \vdots \\ K_N^* \end{bmatrix}, \quad K_0^* := \begin{bmatrix} K_{01}^* \\ \vdots \\ K_{0N}^* \end{bmatrix},$$

$$F_\gamma^* := R_{\gamma W}^{-1} L_{\gamma W},$$

$$R_{\gamma W} := \gamma^2 I_{n_v} - D_p^T W D_p > 0,$$

$$L_{\gamma W} := D^T W + D_p^T W A_p.$$

Then, the following controllers comprise the team-optimal strategy set:

$$u_{0i}^*(t) = K_{0i}^* x(t), \quad i = 1, \ldots, N, \tag{152a}$$

$$u_i^*(t) = K_i^* x(t), \quad i = 1, \ldots, N, \tag{152b}$$

$$v^*(t) = F_\gamma^* x(t) = R_{\gamma W}^{-1} L_{\gamma W} x(t). \tag{152c}$$

Furthermore, the optimal cost bounds are given by

$$J_0(u_1^*, \ldots, u_N^*, v^*, x^0) \leq \mathbb{E}[x^T(0) X^{-1} x(0)] = \mathbf{Tr}[X^{-1}]. \tag{153}$$

As the next step, the conditions of the existence of the follower's strategy set and the incentive are derived. The following LQC problem with the incentive (146) and the worst-case disturbance (152c) is considered:

$$J_i(u_1^\dagger, \ldots, u_{i-1}^\dagger, u_i, u_{i+1}^\dagger, \ldots, u_N^\dagger, \; v^*, \; x^0)$$

$$= \mathbb{E}\left[\int_0^\infty \left\{x^T(t)Q_i x(t) + u_i^T(t)R_{ii}u_i(t)\right\}dt\right], \tag{154}$$

where $u_j^\dagger$, $j \neq i$ means the follower's strategy under the use of the incentive.
In this case, the stochastic system with the incentive (146) is given below.

$$dx(t) = \left[A_{-i\eta}x(t) + B_{\eta i}u_i(t)\right]dt + \bar{A}_p x(t)dw(t), \tag{155a}$$

$$u_{0i}(t) = \eta_{0i}x(t) + \eta_{ii}u_i(t), \tag{155b}$$

where

$$A_{-i\eta}(\theta(t)) := A(\theta(t)) + \sum_{j=1}^N B_{0j}\eta_{0j} + \sum_{j=1,j\neq i}^N B_{\eta j}K_j^* + DF_\gamma^*$$

$$= \sum_{k=1}^M \alpha_k(t)A_{-i\eta k},$$

$$B_{\eta i} := B_i + B_{0i}\eta_{ii},$$

$$\eta_{0i} := K_{0i}^* - \eta_{ii}K_i^*.$$

Theorem 6 ([36]) *Let us consider the stochastic LPV system (144) with multiple decision makers $u_i(t)$ and the deterministic disturbance $v(t)$. For a given attenuation performance level $\gamma > 0$, assume that there exists a solution set for the real symmetric matrices $X_{\eta i} > 0$ and $Y_{\eta i}$ such that the following CCMIs are satisfied: If there exists matrix $X_{\eta i} > 0$ satisfying the LMIs (156):*

$$\begin{bmatrix} \hat{\Xi}_{\eta k}(X_{\eta i}, Y_{\eta i}) & X_{\eta i} & Y_{\eta i}^T & X_{\eta i}\bar{A}_p^T \\ X_{\eta i} & -Q_i^{-1} & 0 & 0 \\ Y_{\eta i} & 0 & -R_{ii}^{-1} & 0 \\ \bar{A}_p X_{\eta i} & 0 & 0 & -X_{\eta i} \end{bmatrix} \leq 0, \; k = 1, \ldots, M, \tag{156}$$

where

$$\hat{\Xi}_{\eta k}(X_{\eta i}, Y_{\eta i}) := A_{-i\eta k}X_{\eta i} + X_{\eta i}A_{-i\eta k}^T + B_{\eta i}Y_{\eta i} + Y_{\eta i}^T B_{\eta i}^T,$$

$$K_{\eta i}^\dagger := Y_{\eta i}X_{\eta i}^{-1}.$$

Then, the following controllers comprise the team-optimal strategy set.

$$u_i^\dagger(t) = K_{\eta i}^\dagger x(t), \; i = 1, \ldots, N. \tag{157}$$

Moreover, we have

$$J_i(u_1^\dagger, \ldots, u_N^\dagger, v^*, x^0) \leq \text{Tr}[X_{\eta i}^{-1}].$$
(158)

Finally, the incentive is given below.

$$u_{0i}(t) = \eta_{0i} x(t) + \eta_{ii}^\dagger u_i(t),$$
(159a)

$$\eta_{0i} = K_{0i}^* - \eta_{ii}^\dagger K_{\eta i}^\dagger, \quad i = 1, \ldots, N,$$
(159b)

$$K_i^* = K_{\eta i}^\dagger, \quad i = 1, \ldots, N.$$
(159c)

It should be noted that the limitations and the disadvantages of the approach is that there does not always exist a solution set of Theorems 5 and 6.

Finally, calculation of η_{ii} is considered. Let us define the following matrix:

$$X_{\eta i}^{-1} = P_{\eta i}.$$
(160)

Through the use of the relation $K_{\eta i} = K_{\eta i}^\dagger = Y_{\eta i} X_\eta^{-1} = Y_{\eta i} P_{\eta i}$ and application of the Schur complement, the original optimization problem with the constraint (156) can be changed to the following optimization problem:

$$\min_{P_{\eta i}, K_{\eta 1}, \ldots, K_{\eta M}} \text{Tr}\,[P_{\eta i}], \text{ s.t. } \Lambda_k(P_{\eta i}, K_{\eta i}) \leq 0,$$
(161)

where

$$\Lambda_k(P_{\eta i}, K_{\eta i}) := P_{\eta i}(A_{-i\eta k} + B_{\eta i} K_{\eta i}) + (A_{-i\eta k} + B_{\eta i} K_{\eta i})^T P_{\eta i}$$
$$+ \bar{A}_p^T P_{\eta i} \bar{A}_p + Q_i + K_{\eta i}^T R_{ii} K_{\eta i}.$$

This optimization problem can be solved using the Karush–Kuhn–Tucker condition. The following Lagrangian L is considered.

$$L = L(P_{\eta i}, K_{\eta i}) = \text{Tr}\,[P_{\eta i}] + \sum_{k=1}^{M} \text{Tr}\,[G_{\eta k} \Lambda_k(P_{\eta i}, K_{\eta i})],$$
(162)

where $G_{\eta k}$ is the symmetric matrix of the Lagrange multiplier. Using the KKT conditions, we obtain

$$G_{\eta k} \Lambda_k(P_{\eta i}, K_{\eta i}) = 0,$$
(163a)

$$G_{\eta k} \geq 0,$$
(163b)

$$\Lambda_k(P_{\eta i}, K_{\eta i}) \leq 0,$$
(163c)

$$\frac{\partial L}{\partial P_{\eta i}} = I_n + \sum_{k=1}^{M} \left[\bar{A}_p G_{\eta k} \bar{A}_p^T + (A_{-i\eta k} + B_{\eta i} K_{\eta i}) G_{\eta k} \right.$$

$$\left. + G_{\eta k} (A_{-i\eta k} + B_{\eta i} K_{\eta i})^T \right] = 0, \tag{163d}$$

$$\frac{1}{2} \cdot \frac{\partial L}{\partial K_{\eta i}} = (B_{\eta i}^T P_{\eta i} + R_{ii} K_{\eta i}) \sum_{k=1}^{M} G_{\eta k} = 0. \tag{163e}$$

It immediately follows that the generalized stochastic Lyapunov equations (163d) have a unique positive definite solution $G_{\eta k} > 0$. Hence, we have $B_{\eta i}^T P_{\eta i} + R_{ii} K_{\eta i}^{\dagger} = 0$. Furthermore, using $R_{ii} > 0$ $K_{\eta i}^{\dagger} = -R_{ii}^{-1} B_{\eta i}^T P_{\eta i}$. On the other hand, using the relation of (159a), we have $K_i^* = K_{\eta i}^{\dagger} \Leftrightarrow -(R_{ii} K_i^* + B_i^T P_{\eta i}) = \eta_{ii}^{\dagger T} B_{0i}^T P_{\eta i}$, $i = 1, \ldots, N$. Finally, if $P_{\eta i} B_{0i}$ is nonsingular, η_{ii} can be computed using the following equation:

$$\eta_{ii}^{\dagger} = -(P_{\eta i} B_{0i})^{-1} (K_i^{*T} R_{ii} + P_{\eta i} B_i), \quad i = 1, \ldots, N. \tag{164}$$

6.4 Numerical Algorithm for Solving CCMIs

In order to construct the team-optimal strategy set (152), we must solve the CCMIs (151). It should be noted that since these matrix inequalities are coupled, the process is very complicated even if an ordinary iterative scheme such as Newton's method is applied. In this section, a numerical algorithm relating to semidefinite programming problems (SDPs) is considered.

First, the numerical algorithm for solving the CCMIs (151) is given.

Step 1 As the first step, choose any weight ρ_i for the cost function (148c) and solve the following SDP.

$$\text{minimize } \alpha^{(0)}, \tag{165}$$

subject to

$$\begin{bmatrix} \Xi_k^{(0)}(X, Y) & X^{(0)} & Y^{(0)T} & X^{(0)} A_{pk}^T \\ X^{(0)} & -Q_0^{-1} & 0 & 0 \\ Y^{(0)} & 0 & -R_c^{-1} & 0 \\ A_{pk} X^{(0)} & 0 & 0 & -X^{(0)} \end{bmatrix} \le 0, \tag{166a}$$

$$\begin{bmatrix} -\alpha^{(0)} & x^T(0) \\ x(0) & -X^{(0)} \end{bmatrix} \le 0, \tag{166b}$$

where $k = 1, \ldots, M$,

$$\Xi_k^{(0)}(X, Y) = A_k X^{(0)} + X^{(0)} A_k^T + B_c Y^{(0)} + Y^{(0)T} B_c^T,$$
$$K^{(0)} = Y^{(0)}[X^{(0)}]^{-1}.$$

Choose any γ and solve $Z^{(0)}$, where

$$Z^{(0)} \bar{A} + \bar{A}^T Z^{(0)} + \bar{A}_p^T Z^{(0)} \bar{A}_p + \gamma^{-2} Z^{(0)} D D^T Z^{(0)} + \bar{E}^T \bar{E} = 0,$$

$$\bar{A} := \frac{1}{N} \sum_{k=1}^{M} A_k, \quad \bar{A}_p := \frac{1}{N} \sum_{k=1}^{M} A_{pk}, \quad \bar{E} := \frac{1}{N} \sum_{k=1}^{M} E_k,$$

$$F^{(0)} = \gamma^{-2} D Z^{(0)}.$$

Step 2 Solve the following SDP.

$$\text{minimize } \alpha^{(p)}, \tag{167}$$

subject to

$$\begin{bmatrix} \Xi_k^{(p)}(X^{(p)}, Y^{(p)}) & X^{(p)} & Y^{(p)T} & X^{(p)} A_{pk}^T \\ X^{(p)} & -Q_0^{-1} & 0 & 0 \\ Y^{(p)} & 0 & -R_c^{-1} & 0 \\ A_{pk} X^{(p)} & 0 & 0 & -X^{(p)} \end{bmatrix} < 0, \tag{168a}$$

$$\begin{bmatrix} -\alpha^{(p)} & x^T(0) \\ x(0) & -X^{(p)} \end{bmatrix} < 0, \tag{168b}$$

where $p = 1, 2, \ldots, k = 1, \ldots, M$,

$$\Xi_k^{(p)}(X^{(p)}, Y^{(p)}) := A_{Fk}^{(p)} X^{(p)} + X^{(p)} A_{Fk}^{(p)T} + B_c Y^{(p)} + Y^{(p)T} B_c^T,$$

$$A_{Fk}^{(p)} := A_k + D F^{(p-1)},$$

$$\begin{bmatrix} K_0^{(p)} \\ K_1^{(p)} \\ \vdots \\ K_N^{(p)} \end{bmatrix} := Y^{(p)}[X^{(p)}]^{-1},$$

$$K_0^{(p)} := \begin{bmatrix} K_{01}^{(p)} \\ \vdots \\ K_{0N}^{(p)} \end{bmatrix}.$$

Step 3 Solve the following SDP.

$$\text{minimize } \mathbf{Tr}\left[x^T(0)Z^{(p)}x(0)\right],$$
(169)

subject to

$$\begin{bmatrix} \boldsymbol{\Theta}_k(W^{(p)}) & W^{(p)}E_k^{(p)T} & W^{(p)}A_{pk}^T \\ E_k^{(p)}W^{(p)} & -I_{\bar{n}_z} & 0 \\ A_{pk}W^{(p)} & 0 & -W^{(p)} \end{bmatrix} < 0,$$
(170)

where $p = 1, 2, \ldots, k = 1, \ldots, M,$

$$\boldsymbol{\Theta}_k(W^{(p)}) := A_{Kk}^{(p)}W^{(p)} + W^{(p)}A_{Kk}^{(p)T} + \gamma^{-2}DD^T,$$

$$A_{Kk}^{(p)} := A_k + \sum_{j=1}^{N}[B_{0j}K_{0j}^{(p)} + B_j K_j^{(p)}],$$

$$E_k^{(p)} := \left[E_k^T \ (G_0 K_0^{(p)})^T \ (G_1 K_1^{(p)})^T \ \cdots \ (G_N K_N^{(p)})^T \right],$$

$$F^{(p)} := \gamma^{-2}D^T[W^{(p)}]^{-1}.$$

Step 4 If the algorithm converges, then $X^{(p)} \to X$, $Y^{(p)} \to Y$ and $W^{(p)} \to W$ as $p \to \infty$. These are the solution of the CCMIs (151), STOP. That is, stop if any norm of the error of difference between the iterative solutions of (168), (170) and the exact solutions of the CCMIs (151) is less than a pre-specified precision. Otherwise, increment $p \to p + 1$ and go to Step 2. If the algorithm does not converge, declare that the algorithm has failed.

Second, the numerical algorithm for solving the LMIs (157) is given.

Step 1 As the first step, choose any initial guess $\eta^{(0)}$ and solve the following SDP.

$$\text{minimize } \beta^{(p)},$$
(171)

subject to

$$\begin{bmatrix} \hat{\boldsymbol{\Xi}}_{\eta k}^{(p)}(X_\eta^{(p)}, Y_\eta^{(p)}) & X_\eta^{(p)} & Y_\eta^{(p)T} & X_\eta^{(p)}A_{pk}^T \\ X_\eta^{(p)} & -Q_\rho^{-1} & 0 & 0 \\ Y_\eta^{(p)} & 0 & -R_\rho^{-1} & 0 \\ A_{pk}X_\eta^{(p)} & 0 & 0 & -X_\eta^{(p)} \end{bmatrix} \le 0,$$
(172a)

$$\begin{bmatrix} -\beta^{(p)} & x^T(0) \\ x(0) & -X_\eta^{(p)} \end{bmatrix} \le 0,$$
(172b)

where $p = 1, 2, \ldots, k = 1, \ldots, M,$

$$\hat{\Xi}_{\eta k}^{(p)}(X_\eta^{(p)}, Y_\eta^{(p)}) := A_{\eta k}^{(p)} X_\eta^{(p)} + X_\eta^{(p)} A_{\eta k}^{(p)T} + B_\eta^{(p)} Y_\eta^{(p)} + Y_\eta^{(p)T} B_\eta^{(p)T},$$

$$A_{\eta k}^{(p)} := A_k + \sum_{j=1}^{N} B_{0j} \eta_{0j}^{(p)} + DF^*,$$

$$B_\eta^{(p)} := \left[B_{1\eta}^{(p)} \cdots B_{N\eta}^{(p)} \right],$$

$$B_{i\eta}^{(p)} := B_i + B_{0i} \eta_{ii}^{(p-1)},$$

$$\eta_{0i}^{(p)} := K_{0i}^* - \eta_{ii}^{(p-1)} K_i^*.$$

Step 2 Compute the following equation:

$$\eta_{ii}^{(p)} = -\left([X_\eta^{(p)}]^{-1} B_{0i}\right)^{-1} \left(\rho_i K_i^T R_i + [X_\eta^{(p)}]^{-1} B_i\right). \tag{173}$$

Step 3 If the algorithm converges, then $X_\eta^{(p)} \to X_\eta$ and $\eta_{ii}^{(p)} \to \eta_{ii}$ as $p \to \infty$. These are the solution of the LMIs (157), STOP. Otherwise, increment $p \to p+1$ and go to Step 2. If the algorithm does not converge, declare that the algorithm has failed.

6.5 Numerical Example

A simple numerical example is investigated to demonstrate the efficiency of our proposed three strategies. The system of matrices is as follows:

$$A_1 = \begin{bmatrix} -1.36 & 2 \\ -1 & -1.55 \end{bmatrix}, \; A_{p1} = 0.1 A_1,$$

$$A_2 = \begin{bmatrix} -1.36 & 0 \\ -1 & -1.55 \end{bmatrix}, \; A_{p2} = 0.1 A_2,$$

$$\alpha_1(t) = \sin^2 t, \; \alpha_2(t) = \cos^2 t,$$

$$B_{01} = \begin{bmatrix} 1 & 0 \\ 1 & 4.15 \end{bmatrix}, \; B_{02} = \begin{bmatrix} 1 & 0 \\ 0.2 & 1.32 \end{bmatrix},$$

$$B_1 = \begin{bmatrix} 1 & 1 \\ 1.2 & 2.65 \end{bmatrix}, \; B_2 = \begin{bmatrix} 1 & 0 \\ 0.4 & 1.32 \end{bmatrix},$$

$$D = \begin{bmatrix} 0.1 \\ 0.2 \end{bmatrix}, \; E_1 = I_2, \; E_2 = \begin{bmatrix} 2 & 0 \\ 0 & 1 \end{bmatrix}, \; \cdot$$

$$G_0 = \begin{bmatrix} I_2 \ I_2 \end{bmatrix}, \ G_1 = G_2 = I_2,$$

$$Q_1 = \begin{bmatrix} 2 & 0 \\ 0 & 0.5 \end{bmatrix}, \ Q_2 = \begin{bmatrix} 1 & 0 \\ 0 & 2 \end{bmatrix},$$

$$R_{001} = 2I_2, \ R_{002} = 4I_2, \ R_{01} = 3I_2, \ R_{02} = 2I_2,$$

$$R_1 = 2I_2, \ R_2 = 3I_2.$$

The disturbance attenuation level γ is chosen as $\gamma = 5$. The CCMIs (151) are solved by using the algorithm of the previous subsection. The strategy set that attains the Pareto optimal solution with the H_∞ constraint is given below.

$$K_1^* = \begin{bmatrix} -1.0225\mathrm{e}\text{-}1 & -1.5130\mathrm{e}\text{-}1 \\ -8.2875\mathrm{e}\text{-}2 & -3.5027\mathrm{e}\text{-}1 \end{bmatrix},$$

$$K_2^* = \begin{bmatrix} -1.6941\mathrm{e}\text{-}1 & -6.2288\mathrm{e}\text{-}2 \\ 2.6458\mathrm{e}\text{-}2 & -2.7170\mathrm{e}\text{-}1 \end{bmatrix},$$

$$K_{01}^* = \begin{bmatrix} -1.5739\mathrm{e}\text{-}1 & -1.8579\mathrm{e}\text{-}1 \\ 8.3182\mathrm{e}\text{-}2 & -8.5420\mathrm{e}\text{-}1 \end{bmatrix},$$

$$K_{02}^* = \begin{bmatrix} -8.6710\mathrm{e}\text{-}2 & -1.0561\mathrm{e}\text{-}2 \\ 1.3229\mathrm{e}\text{-}2 & -1.3585\mathrm{e}\text{-}1 \end{bmatrix},$$

$$F^* = \begin{bmatrix} 3.7110\mathrm{e}\text{-}3 & 1.3905\mathrm{e}\text{-}3 \end{bmatrix}.$$

The incentive (164) and the related matrices are given below.

$$\eta_{11} = \begin{bmatrix} -7.4267\mathrm{e}\text{-}1 & -7.6536\mathrm{e}\text{-}1 \\ 5.7613\mathrm{e}\text{-}2 & -6.9769\mathrm{e}\text{-}2 \end{bmatrix},$$

$$\eta_{22} = \begin{bmatrix} -3.9285\mathrm{e}\text{-}1 & -4.6462\mathrm{e}\text{-}2 \\ 9.8427\mathrm{e}\text{-}2 & 4.0154\mathrm{e}\text{-}1 \end{bmatrix},$$

$$\eta_{01} = \begin{bmatrix} -3.3440\mathrm{e}\text{-}1 & -4.8297\mathrm{e}\text{-}1 \\ 6.3514\mathrm{e}\text{-}2 & -7.3710\mathrm{e}\text{-}1 \end{bmatrix},$$

$$\eta_{02} = \begin{bmatrix} -1.6417\mathrm{e}\text{-}1 & -4.1802\mathrm{e}\text{-}2 \\ 2.0032\mathrm{e}\text{-}2 & -1.7314\mathrm{e}\text{-}2 \end{bmatrix},$$

$$P_\eta = \begin{bmatrix} 4.5636\mathrm{e}\text{-}1 & -5.0334\mathrm{e}\text{-}3 \\ -5.0334\mathrm{e}\text{-}3 & 1.8747\mathrm{e}\text{-}1 \end{bmatrix}.$$

The proposed SDP algorithm converges to the required solution with an accuracy of 1.0e-7 after eight iterations. The algorithm based on SDPs is easy to implement in MATLAB. However, it should be noted that there is no proof for the convergence of the SDP algorithm.

Finally, after announcing this incentive strategy (160), the strategy set of the followers can be computed. Indeed, it can be observed that the matrix gain K_i^* equals $K_{\eta i}^{\dagger}$, $i = 1, \ldots, N$ after the followers have made a decision. Namely, it can be confirmed that the followers are induced to the team-optimal solution eventually.

7 Conclusion

The incentive Stackelberg game under the H_∞ constraint for the stochastic systems has been reviewed. Unlike the existing deterministic Stackelberg games [13–26], the stochastic incentive Stackelberg game has been studied for the first time in this work. Table 1 summarizes the recent contributions by the author. It should be noted that the results in Sect. 5 are novel because the SOF incentive Stackelberg strategy with H_∞ constraint is investigated for the first time. In this survey, it has been shown that the proposed incentive Stackelberg strategy set can be computed by solving a set of cross-coupled stochastic Riccati-type equations or linear matrix inequalities. The leader's team-optimal solution with the H_∞ constraint can be achieved eventually under this design. As an important feature of the incentive Stackelberg games, it is worth noting that multiple followers are subjected to Nash equilibrium or Pareto

Table 1 Recent results in Stochastic Incentive Stackelberg games

Reference	Model	Player	Feedback type	Robustness
Journal [27]	CSS	ML	SF	
Journal [35]	CSS	MF	SF	H_∞ control
Journal [31]	DSS	MF	SF	H_∞ control
Journal [33]	CMJSS	MLF	SF	
Journal [34]	DMJSS	SLF	SOF	H_∞ control
Proc. [28]	DDS	MF	SF	H_∞ control
Proc. [29]	DSS	SLF	SF	H_∞ control
Proc. [30]	CMJSS	MLF	SF	
Proc. [32]	CMJSS	MF	SF	H_∞ control
Proc. [37]	CMJSS	MLF	SF	H_∞ control
Proc. [36]	SLPV	MF	SF	H_∞ control
Proc. [38]	SLPV	MF	SF	H_∞ control

CSS: continuous-time stochastic system, DSS: discrete-time stochastic system, DDS: discrete-time deterministic system, CMJSS: continuous-time Markov jump stochastic system, DMJSS: discrete-time Markov jump stochastic system, SLPV: stochastic linear parameter varying systems, ML: multiple-leader, MF: multiple-follower, SLF: single-leader-follower, MLF: multiple leader-follower, SF: state feedback, SOF: static output feedback
It should be noted that in [36], the followers consider Nash equilibrium solutions, and in [38], the followers consider Pareto optimal solutions. It should also be noted that the IEEE Control Systems Letters (L-CSS) offers the opportunity for authors to not only publish a paper in the journal but also to present the same paper at the flagship conference of the IEEE Control Systems Society: the IEEE Conference on Decision and Control (CDC).

optimal conditions via the imposed incentive, whereas the leader achieves a team-optimal solution under an H_∞ constraint.

The incentive Stackelberg game is an important long-standing research area; however, there are still unsolved problems. First, the existence condition of the incentive needs to be discussed more deeply. For example, if the decision maker's input u_{0i} is a set of multi-variables and the related incentive parameter η_{1i} is a scalar case, it is easy to observe that there exists no solution. Furthermore, to the best of our knowledge, incentive Stackelberg games for stochastic systems in cases other than LQ have not been investigated. Consequently, proof of the existence and uniqueness of the incentive Stackelberg strategy set in such systems will be challenging. Thus, it is expected that these problems will be addressed in future studies.

References

1. T. Basar and G. J. Olsder, *Dynamic Noncooperative Game Theory*, Philadelphia: SIAM Series in Classics in Applied Mathematics, 1999.
2. V. R. Saksena and J. B. Cruz, Jr, Optimal and near-optimal incentive strategies in the hierarchical control of Markov chains, *Automatica*, vol. 21, no. 2, pp. 181–191, 1985.
3. K. B. Kim, A. Tang and S. H. Low, "A stabilizing AQM based on virtual queue dynamics in supporting TCP with arbitrary delays," in *Proc. 42nd IEEE Conf. Decision and Control*, Maui, HI, December 2003, pp. 3665–3670.
4. R. Kicsiny, Solution for a class of closed-loop leader-follower games with convexity conditions on the payoffs, *Annals of Operations Research*, vol. 253, no. 1, pp. 405–429, 2017.
5. M. Yu and S. H. Hong, Supply-demand balancing for power management in smart grid: A Stackelberg game approach, *Applied Energy*, vol. 164, pp. 702–710, 2016.
6. C. I. Chen and J. B. Cruz, Jr., Stackelberg solution for two-person games with biased information patterns, *IEEE Trans. Automatic Control*, vol. 17, no. 6, pp. 791–798, Dec. 1972.
7. J. V. Medanic, Closed-loop Stackelberg strategies in linear-quadratic problems, *IEEE Trans. Automatic Control*, vol. 23, no. 4, pp. 632–637, Aug. 1978.
8. M. Jungers, E. Trelat and H. Abou-Kandil, Min-max and min-min Stackelberg strategies with closed-loop information structure, *J. Dynamical and Control Systems*, vol. 17, no. 3, pp. 387–425, 2011.
9. A. Bensoussan, S. Chen and S. P. Sethi, The maximum principle for global solutions of stochastic Stackelberg differential games, *SIAM Control and Optimization*, vol. 53, no. 4, pp. 1956–1981, 2015.
10. J. Xu and H. Zhang, Sufficient and necessary open-loop Stackelberg strategy for two-player game with time delay, *IEEE Trans. Cybernetics*, vol. 46, no. 2, pp. 438–449, Feb. 2016.
11. H. Mukaidani and H. Xu, Stackelberg strategies for stochastic systems with multiple followers, *Automatica*, vol. 53, pp. 53–59, 2015.
12. H. Mukaidani and H. Xu, Infinite-horizon linear-quadratic Stackelberg games for discrete-time stochastic systems, *Automatica*, vol. 76, no. 5, pp. 301–308, 2017.
13. Y. C. Ho, Peter B. Luh and G. J. Olsder, A control-theoretic view on incentives, *Automatica*, vol. 18, no. 2, pp. 167–179, 1982.
14. T. Basar and H. Selbuz, Closed-loop Stackelberg strategies with applications in the optimal control of multilevel systems, *IEEE Trans. Automatic Control*, vol. 24, no. 2, pp. 166–178, April 1979.
15. B. Tolwinski, Closed-loop Stackelberg solution to multi-stage linear quadratic game, *J. Optimization Theory and Applications*, vol. 34, no. 3, pp. 485–501, 1981.

16. Y. P. Zheng and T. Basar, Existence and derivations of optimal affine incentive schemes for Stackelberg games with partial information: A geometric approach, *Int. J. Control*, vol. 35, no. 6, pp. 997–1011, 1982.
17. Y. P. Zheng, T. Basar and J. B. Cruz Jr., Stackelberg strategies and incentives in multiperson deterministic decision problems, *IEEE Trans. Systems, Man, and Cybernetics*, vol. 14, no. 1, pp. 10–24, Jan. 1984.
18. T. Basar and G. J. Olsder, Team-optimal closed-loop Stackelberg strategies in hierarchical control problems, *Automatica*, vol. 16, no. 3, pp. 409–414, 1980.
19. P. B. Luh, S. C. Chang and T.S. Chang, Solutions and properties of multi-stage Stackelberg games, *Automatica*, vol. 20, no. 2, pp. 251–256, 1984.
20. T.S. Chang and P. B. Luh, Derivation of necessary and sufficient conditions for single-state Stackelberg games via the inducible region concept, *IEEE Trans. Automatic Control*, vol. 29, no. 1, pp. 63–66, 1984.
21. P. B. Luh, T. S. Chang and T. Ning, Three-level Stackelberg decision problems, *IEEE Trans. Automatic Control*, vol. 29, no. 3, pp. 280–282, 1984.
22. K. Mizukami and H. Wu, Two-level incentive Stackelberg strategies in LQ differential games with two noncooperative leaders and one follower, Trans. SICE, vol. 23, no. 6, pp. 625–632, 1987.
23. K. Mizukami and H. Wu, Incentive Stackelberg strategies in linear quadratic differential games with two noncooperative followers, Modelling And Methodology In Social And Economic Systems System Modelling and Optimization Volume 113 of the series Lecture Notes in Control and Information Sciences, pp. 436–445, Berlin Heidelberg: Springer, 1988.
24. T. Ishida and E. Shimemura, Three-level incentive strategies in differential games, *Int. J. Control*, vol. 38, no. 6, 1983, pp. 1135–1148.
25. T. Basar, Equilibrium strategies in dynamic games with multilevel of hierarchy, *Automatica*, vol. 17, no. 5, pp. 749–754, 1981.
26. M. Li, J. B. Cruz Jr., M. A. Simaan, An approach to discrete-time incentive feedback Stackelberg games, *IEEE Trans. Systems, Man, and Cybernetics*, vol. 32, no. 4, pp. 10–24, July 2002.
27. H. Mukaidani, Infinite-horizon team-optimal incentive Stackelberg games for linear stochastic systems, *IEICE Trans. Fundamentals of Electronics, Communications and Computer Sciences*, vol. E99-A, no. 9, pp. 1721–1725, 2016.
28. M. Ahmed and H. Mukaidani, "H_∞-constrained incentive Stackelberg game for discrete-time systems with multiple non-cooperative followers," in *Proc. 6th IFAC Workshop on Distributed Estimation and Control in Networked Systems*, Tokyo, Japan, September 2016, *IFAC-PapersOnLine*, 49-22, pp. 262–267.
29. H. Mukaidani, M. Ahmed, T. Shima and H. Xu, "H_∞ constraint incentive Stackelberg game for discrete-time stochastic systems," in *Proc. American Control Conf.*, Seattle, WA, May 2017, pp. 5257–5262.
30. M. Ahmed, H. Mukaidani and T. Shima, "Infinite-horizon multi-leader-follower incentive Stackelberg games for linear stochastic systems with H_∞ constraint," in *SICE Annual Conf.*, Kanazawa, Japan 2017, pp. 1202–1207.
31. M. Ahmed, H. Mukaidani and T. Shima, H_∞-constrained incentive Stackelberg games for discrete-time stochastic systems with multiple followers, *IET Control Theory and Applications*, vol. 11, no. 15, pp. 2475–2485, 2017.
32. H. Mukaidani and T. Shima, M. Unno, H. Xu and V. Dragan, "Team-optimal incentive Stackelberg strategies for Markov jump linear stochastic systems with H_∞ constraint," in *Proc. 20th IFAC World Congress*, Toulouse, France, July 2017, *IFAC-PapersOnLine*, 50-1, pp. 3780–3785.

33. H. Mukaidani, H. Xu and V. Dragan, A stochastic multiple-leader follower incentive Stackelberg strategy for Markov jump linear systems, *IEEE Control Systems Letters*, vol. 1, no. 2, pp. 250–255, 2017. (See also 56th IEEE Conf. Decision and Control, pp. 3688–3693, Melbourne, Australia, December 2017.)

34. H. Mukaidani, H. Xu and V. Dragan, Static output-feedback incentive Stackelberg game for discrete-time Markov jump linear stochastic systems with external disturbance, *IEEE Control Systems Letters*, vol. 2, no. 4, pp. 701–706, 2018. (See also 57th IEEE Conf. Decision and Control, Miami, FL, December 2018.)

35. H. Mukaidani and H. Xu, Incentive Stackelberg games for stochastic linear systems with H_∞ constraint, *IEEE Trans. Cybernetics*, vol. 49, no. 4, pp. 1463–1474, 2019.

36. K. Kawakami, H. Mukaidani, H. Xu and Y. Tanaka, "Incentive Stackelberg-Nash strategy with disturbance attenuation for stochastic LPV systems," in *Proc. the 2018 IEEE Int. Conf. Systems, Man, and Cybernetics*, Miyazaki, Japan, October 2018, pp. 3940–3945.

37. H. Mukaidani, H. Xu, T. Shima and M. Ahmed, "Multi-leader-follower incentive Stackelberg game for infinite-horizon Markov jump linear stochastic systems with H_∞ constraint," in *Proc. the 2018 IEEE Int. Conf. Systems, Man, and Cybernetics*, Miyazaki, Japan, October 2018, pp. 3946–3953.

38. H. Mukaidani and Hua Xu, "Robust incentive Stackelberg games for stochastic LPV systems," in *Proc. 57th IEEE Conf. Decision and Control*, Miami, FL, December 2018, pp. 1059–1064.

39. B. S. Chen and W. Zhang. Stochastic H_2/H_∞ control with state-dependent noise, *IEEE Trans. Automatic Control*, vol. 49, no. 1, pp. 45–57, Jan. 2004.

40. W. Zhang and B. -S. Chen, On stabilizability and exact observability of stochastic systems with their applications, *Automatica*, vol. 40, no. 1, pp. 87–94, 2004.

41. J. C. Engwerda, *LQ Dynamic Optimization and Differential Games*, Chichester: John Wiley and Sons, 2005.

42. H. Mukaidani, Soft-constrained stochastic Nash games for weakly coupled large-scale systems, *Automatica*, vol. 45, no. 5, pp. 1272–1279, 2009.

43. Z. Gajic and M. Qureshi, *Lyapunov Matrix Equation in System Stability and Control*, San Diego, Academic Press, Mathematics in Science and Engineering Series, 1995.

44. M. Sagara, H. Mukaidani and T. Yamamoto, Numerical solution of stochastic Nash games with state-dependent noise for weakly coupled large-scale systems, *Applied Mathematics and Computation*, vol. 197, no. 2, pp 844–857, 2008.

45. V. Dragan and T. Morozan, The linear quadratic optimization problems for a class of linear stochastic systems with multiplicative white noise and Markovian jumping, *IEEE Trans. Automatic Control*, vol. 49, no. 5, pp. 665–675, 2004.

46. Y. Huang, W. Zhang and G. Feng, Infinite-horizon H_2/H_∞ control for stochastic systems with Markovian jumps, *Automatica*, vol. 44, no. 3, pp. 857–863, 2008.

47. X. Li, X. Y. Zhou and M. A. Rami, Indefinite stochastic linear quadratic control with Markovian jumps in infinite time horizon, *J. Global Optimization*, vol. 27, no. 2–3, pp. 149–175, 2003.

48. H. Mukaidani, "Gain-scheduled H_∞ constraint Pareto optimal strategy for stochastic LPV systems with multiple decision makers," in *American Control Conf.*, Seattle, WA, May 2017, pp. 1097–1102.

49. W. Zhang and G. Feng, Nonlinear stochastic H_2/H_∞ control with (x, u, v)-dependent noise: Infinite-horizon case, *IEEE Trans. Automatic Control*, vol. 53, no. 5, pp 1323–1328, 2008.

50. M. A. Rami and X. Y. Zhou, Linear matrix inequalities, Riccati equations, and indefinite stochastic linear quadratic controls, *IEEE Trans. Automatic Control*, vol. 45, no. 6, pp 1131–1143, 2000.

51. C. -C. Ku and C. -I. Wu, Gain-scheduled H_∞ control for linear parameter varying stochastic systems, *J. Dynamic Systems, Measurement, and Control*, vol. 137, no. 11, 2015, 111012–1.

Social and Private Interests Coordination Engines in Resource Allocation: System Compatibility, Corruption, and Regional Development

Olga I. Gorbaneva and Gennady A. Ougolnitsky

Abstract This paper analyzes conditions of the system compatibility in the static game-theoretic models of resource allocation between social and private activities. We describe administrative and economic control mechanisms providing system compatibility and formalize them as static Stackelberg and inverse Stackelberg games. Descriptive and normative approaches to the modeling of corruption in resource allocation in the hierarchical control systems are proposed and implemented. Applications to the problems of regional development are outlined.

Keywords Social and private interests · SPICE-models · Social welfare functions · Price of anarchy · System consistensy · Administrative and economical mechanisms · Descriptive and normative approaches.

1 Introduction

The problem of coordination of private and social interests plays a key role in the analysis of economic relations. There are several directions of research in this domain considering the resource allocation.

Economics of public goods [6, 7] studies their production and the respective cost distribution. A pure public good is consumed by all participants of a system independently of their contribution to its production. This differs it from a pure private good which can be distributed and sold. Samuelson [26] defined the pure public good as providing non-excludable and non-competitive profits to all members of a society on the local, regional, or national level. Warr [27] proved that if private agents participate in a voluntary production of a public good then this production does not depend on the redistribution of the income. Bernheim and

O. I. Gorbaneva (✉) · G. A. Ougolnitsky
J.I. Vorovich Institute of Mathematics, Mechanics and Computer Sciences, Southern Federal University, Rostov-on-Don, Russian Federation
e-mail: oigorbaneva@sfedu.ru; gaugolnickiy@sfedu.ru

© Springer Nature Switzerland AG 2020
D. Yeung et al. (eds.), *Frontiers in Games and Dynamic Games*, Annals of the International Society of Dynamic Games 16,
https://doi.org/10.1007/978-3-030-39789-0_4

119

Bagwell [5] and Kemp [18] extended this result for the case of several public goods. Bergstrom et al. [2] improved the result as follows. A small redistribution of material goods between investing consumers does not change an equilibrium quantity of the public good but an essential redistribution changes the set of investors and the equilibrium distribution as well [2–4]. Some recent results are presented in [9, 17].

It is necessary to notice a seminal paper by Germeier and Vatel [10] which is improperly ignored by the majority of publications in this domain. In this paper a game in normal form is studied where the payoff functions are convolutions by minimum of two terms. The first one represents a private interest of the player and the second one represents the same common interest of all players. It is proved that in natural conditions this game has a Pareto-optimal Nash equilibrium. This line of research was continued, for example, in [19].

Control mechanisms of the resource allocation are analyzed in the theory of control in organizational systems [8, 20, 21]. A set of the strategy-proof mechanisms of resource allocation is proposed.

An important place belongs to the phenomenon of corruption in resource allocation between purpose (social) and non-purpose (private) interests in the hierarchical systems [16].

At last, in regional development the issues of resource allocation arise in the context of relations between the federal and regional (or regional and local) administrations as well as in the trans-frontier unions and state-private partnerships [25].

Our approach to the problem is based on the following assertions and assumptions [11–13].

1. There are n agents. Each of them allocates his resource between his private activity and the production of a social good.
2. Then the social good is divided between the agents in given or controlled shares. This defines the difference between the social good and the pure public good.
3. The payoff function of each agent consists of two summands. The first one represents the agent's income from his private activity and the second one is his share in the social good.
4. A concept of system compatibility is introduced which characterizes the degree of coordination of private and social interests in the system quantitatively.
5. Real economic organizations are system compatible very rarely. So, a special agent (Center) is introduced which represents the social interests (maximizes the social welfare) and tries to ensure the system compatibility.
6. The Center (social planner) controls the agents by two methods. First, she can restrict the amount of resource assigned by agents to their private activity (administrative mechanisms, compulsion). Second, she can determine their shares in the social good (economic mechanisms, impulsion). From the mathematical point of view, these control mechanisms are formalized as Stackelberg games or inverse Stackelberg games [1, 22, 23]. We call such models Social and Private Interests Coordination Engines (SPICE-models).
7. We distinguish administrative and economic corruption [14]. In the case of administrative corruption, the restrictions on the set of feasible strategies of

an agent can be weakened in exchange for a bribe. In the case of economic corruption, a bribe permits to increase the share of an agent in the social good. In both cases corruption is treated as a feedback on the control bribe variable that creates an inverse Stackelberg game.

The rest of the paper is organized as follows. In Sect. 2 the SPICE-models for two and several players are defined and the conditions of system compatibility are analyzed in the cases of independent players, hierarchy, and cooperation. In Sect. 3 economic and administrative mechanisms of system compatibility are introduced. Section 4 is dedicated to the economic and administrative corruption in resource allocation. In Sect. 5 the SPICE-models are applied to the problems of regional development. The economic, administrative, and resource allocation control mechanisms are considered. The last section concludes.

2 SPICE-Models and Conditions of System Compatibility

In the basic setup, SPICE-models are static games in normal form which formalize the following situation. There are n agents, each of them allocates his resource between his private activity and the production of a social good. Then the social good is divided between the agents in given or controlled shares. This defines the difference between the social good and the pure public good. The payoff function of each agent consists of two summands. The first one represents the agent's income from his private activity and the second one is his share in the social good. A concept of system compatibility is introduced which characterizes the degree of coordination of private and social interests in the system quantitatively.

The games are considered for different information structures. First, we consider a standard setup with independent players who choose their control variables independently and simultaneously, and the solution is a Nash equilibrium. Second, a hierarchical setup is considered where a solution in the sense of Stackelberg is searched. Third, the case of cooperation is analyzed where a team solution which is Pareto optimal is attained.

The cases of SPICE-models with two and several agents are considered sequentially.

2.1 SPICE-Models with Two Agents

The model has the form

$$g_1(u_1, u_2) = p_1(r_1 - u_1) + s_1(u_1, u_2)c(u_1, u_2) \to \max_{u_1},$$

$$g_2(u_1, u_2) = p_2(u_1, r_2 - u_2) + s_2(u_1, u_2)c(u_1, u_2) \to \max_{u_2}.$$

There are evident restrictions $0 \le u_i \le r_i$ and conditions for the functions p_i, b, c

$$p_i \ge 0, \frac{\partial p_i}{\partial u_i} \le 0, \frac{\partial p_i}{\partial u_{j \ne i}} \ge 0, s_i \ge 0, \frac{\partial s_i}{\partial u_i} \ge 0, \frac{\partial c}{\partial u_i} \ge 0, i = 1, 2.$$

Here r_i is a number of resource of the i-th agent, u_i—a share of this number assigned for the production of a social good (the agent's control), p_i—a function of the income from private activity of the i-th agent, s_i is a share of the agent in the social good; c is a production function of the social good.

Unlike the Germeier–Vatel model [10] we use a linear convolution of the functions which represent private and social interests. In fact, they are production functions widely used in mathematical economics. Namely, we use power functions $p_i(\cdot)$ and $c(\cdot)$ with a positive exponent less or equal than one. The following ways of distribution of the social good between the agents are considered:

1. a constant one $s_1 = s$, $s_2 = 1 - s$;
2. a proportional one $s_1 = \frac{u_1}{u_1 + u_2}$, $s_2 = \frac{u_2}{u_1 + u_2}$.

2.1.1   Independence

An economic system consists of two equal agents A_1 and A_2 having numbers of resources r_i, $i = 1, 2$. A share u_i is assigned for the production of social good, the rest for a private activity. The structure of the system is presented in Fig. 1. The model is a game in normal form in which a Nash equilibrium is searched:

$$g_1(u_1, u_2) = p_1(r_1 - u_1) + s_1(u_1, u_2)c(u_1, u_2) \to \max_{u_1},$$

$$g_2(u_1, u_2) = p_2(r_2 - u_2) + s_2(u_1, u_2)c(u_1, u_2) \to \max_{u_2},$$

with constraints $0 \le u_i \le r_i$.

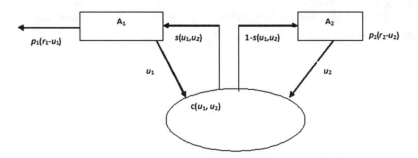

Fig. 1 The structure of the modeled system in the case of independence

The functions p_i and c are production functions: $p_1 = p_1(r_1 - u_1)$, $p_2 = p_2(r_2 - u_2)$, $c = c(u_1 + u_2)$. Here we suppose that the production of a social good depends on summary investments. The set of Nash equilibria is found for power production functions [11].

2.1.2 Hierarchy

Now suppose that two agents are hierarchically subordinated. Center A_1 has a resource r. She transfers a part u_1 of it to the agent A_2 for the production of a social good using the rest for a private activity. In turn, A_1 assigns a part u_2 of the received amount u_1 for the production of the social good using the rest for his private activity. Both agents receive a share from the produced social good (Fig. 2). The model is a two-player Stackelberg game in the form

$$g_1(u_1, u_2) = p_1(r(1 - u_1)) + sc(ru_1u_2) \to \max_{u_1},$$

$$g_2(u_1, u_2) = p_2(ru_1(1 - u_2)) + (1 - s)c(ru_1u_2) \to \max_{u_2},$$

with constraints $0 \le u_i \le 1$, $i = 1, 2$. The set of Stackelberg equilibria is found for power production functions. It is shown that from the point of view of the production of a social good independence is better than hierarchy [11].

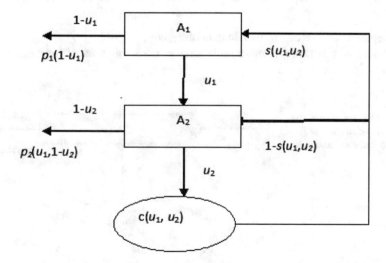

Fig. 2 The structure of the modeled system in the case of hierarchy

2.1.3 Cooperation

Here we consider two subcases: independent and hierarchical cooperation. In the case of independent cooperation, two equal agents A_1 and A_2 have amounts of resource r_i and assign a part u_i for the production of a social good whether the rest is used for a private activity. These agents form a coalition, join their resources, have the summary payoff function, choose their strategies, and then divide the total payoff together (Fig. 3). The model is an optimization problem that leads to a team solution which is Pareto-optimal. In the case of independent cooperation, two agents are initially hierarchically subordinated. Center A_1 has a resource r. She transfers a part u_1 of it to the agent A_2 for the production of a social good using the rest for a private activity. In turn, A_1 assigns a part u_2 of the received amount u_1 for the production of the social good using the rest for his private activity. Then the agents form a coalition, join their resources, have the summary payoff function, choose their strategies, and then divide the total payoff together.

The model is an optimization problem

$$g_0(u_1, u_2) = p_1(r(1 - u_1)) + p_2(ru_1(1 - u_2)) + c(ru_1u_2) \to \max_{u_1, u_2},$$

with constraints $0 \le u_i \le 1$ that leads to a team solution. The results are presented in [11]. The influence of the information structure to resource allocation is shown in Tables 1 and 2. Thus, the tables show that from the point of view of a social planner:

1. independence is more preferable than hierarchy;
2. creation of coalitions is more preferable in both subcases of independent and hierarchical coalitions.

Fig. 3 The structure of the modeled system in the case of independent cooperation

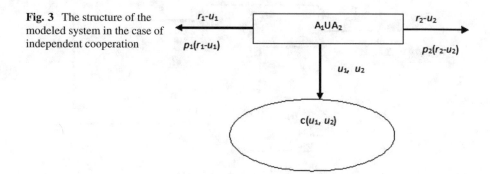

Table 1 Structural influence

Aspect of comparison	Independence	Hierarchy
Amount of the resources assigned for the production of a social good (variable u_i)	Greater	Less
Probability of complete individualism (all resources of both agents are assigned for their private activities)	Less	Greater
Probability of partial individualism (all resources of at least one agent are assigned for his private activity)	Less	Greater

Table 2 Cooperation influence

Aspect of comparison	Non-coalitional system	Coalitional system
Amount of the resources assigned for the production of a social good (variable u_i)	Less	Greater
Probability of complete individualism (all resources of both agents are assigned for their private activities)	Greater	Less
Probability of partial individualism (all resources of at least one agent are assigned for his private activity)	Greater	Less

2.2 SPICE-Models with Several Agents

Now consider SPICE-models with several agents. The respective game in normal form is

$$g_i(u_1, \ldots, u_n) = p_i(r_i - u_i) + s_i c(u_1, \ldots, u_n) \to \max \qquad (1)$$

$$0 \le u_i \le r_i, r_i \ge 0, s_i \ge 0, \sum_{j=1}^{n} s_j = \begin{cases} 1, & \exists i : s_i > 0 \\ 0, & \forall i : s_i = 0, \end{cases}, i = 1, \ldots, n. \qquad (2)$$

Here $N = \{1, \ldots, n\}$—a set of players; $U_i = [0, r_i]$—a set of feasible strategies of the i-th player; r_i—a number of resources of the i-th player; $g_i(u_1, \ldots, u_n)$—a payoff function of the i-th player; $g_i : U \to R, U = U_1 \times \ldots \times U_n$; $p_i(r_i - u_i)$—a function of income of the i-th player from his private activity; $c(u_1, \ldots, u_n)$—a function of the production of a social good; s_i—a share of the social good allocated to the player i; $s_i c(u_1, \ldots, u_n)$—the i-th player's payoff from using the social good.

Function c increases monotonically by all u_i, $c(0, \ldots, 0) = 0$, functions p_i increase monotonically by $(r_i - u_i)$ and decrease monotonically by u_i, $p_i(0) = 0$ (when $u_i = r_i$); if $s_i > 0$, then $u_i > 0$.

Introduce a social welfare function

$$g(u_1, \ldots, u_n) = \sum_{j=1}^{n} g_j(u_1, \ldots, u_n) = \sum_{j=1}^{n} p_j(r_j - u_j) + c(u_1, \ldots, u_n). \qquad (3)$$

Denote $NE = \{u_{(1)}^{NE}, \ldots, u_{(k)}^{NE}\}$—a set of Nash equilibria in the game (1)–(2), $g_{\min}^{NE} = \min\{g(u_{(1)}^{NE}), \ldots, g(u_{(k)}^{NE})\}$, $g_{\max} = \max_{u \in U} g(u) = g(u^{\max})$. Then the price of anarchy in the model (1)–(2) is equal to

$$PA = \frac{g_{\min}^{NE}}{g_{\max}}. \tag{4}$$

It is evident that $PA \leq 1$. If it is close to one, then the efficiency of equilibria is high and the need in coordination in the model (1)–(2) is small or absent at all (when $PA = 1$); the less PA, the less the need in coordination.

System Compatibility Introduce the following

Definition 1 The model (1)–(3) is system compatible if $\exists u^{NE} \in NE : u^{NE} = u^{\max}$.

As the analysis shows, in the majority of cases $PA < 1$, i.e. an egoistic behavior of the agents leads to non-efficient equilibria. It is evident that if $PA = 1$ then the model is system compatible. Introduce the following sets:

$I = \{i : u_i = 0\}$—a set of individualists (they assign all resources for their private activities).

$C = \{i : u_i = r_i\}$—a set of collectivists (they assign all resources for the production of a social good). Then the following main result holds.

Theorem 1 ([13]) *If $n \geq 2$ and $p_i(0) = 0$, $c(0) = 0$; functions c, p_i are increasing and concave, then the system compatibility holds only if there is a partition of the set of agents on two classes: individualists and collectivists ($N = I \cup C$, $I \cap C = \emptyset$).*

3 Control Mechanisms of System Compatibility

The condition of system compatibility is satisfied rarely by itself, and it is necessary to use control mechanisms to ensure it. Suppose that maximization of the social welfare is the objective of a specific agent (Center, Principal, social planner, mechanism designer) who can exert influence to the other agents to achieve this goal. The Center can exert influence to the sets of feasible controls (administrative mechanism, compulsion) or payoff functions of the other agents (economic mechanism, impulsion). Denote the first possibility by $U_i = U_i(q_i)$ and the second one by $g_i = g_i(p_i, u_i)$. Both methods of control may not use or use a feedback by control. In the first case—a Stackelberg game (a Germeier game of the type $\Gamma 1$) arises, in the second case—an inverse Stackelberg game (a Germeier game of the type $\Gamma 2$). Thus, there are four types of the mechanisms which are presented in Table 3 [24].

Definition 2 A control mechanism in the model (1)–(3) is system compatible if the best response of the agents to this mechanism makes the model system compatible.

Table 3 Control mechanisms

The Center's influence	Without a feedback (Stackelberg game, $\Gamma 1$)	With a feedback (inverse Stackelberg game, $\Gamma 2$)
To the sets of feasible controls of the agents (administrative, compulsion)	Administrative impact without a feedback $q_i = const$	Administrative impact with a feedback $q_i = q_i(u_i)$
To their payoff functions (economic, impulsion)	Economic impact without a feedback $p_i = const$	Economic impact with a feedback $p_i = p_i(u_i)$

Table 4 Control mechanisms in SPICE-models

The Center's influence	Without a feedback (Stackelberg game, $\Gamma 1$)	With a feedback (inverse Stackelberg game, $\Gamma 2$)
To the sets of feasible controls of the agents (administrative, compulsion)	Administrative impact without a feedback $\tilde{q}_i \leq u_i \leq \bar{q}_i, \tilde{q}_i, \bar{q}_i = const$	Administrative impact with a feedback $\tilde{q}_i(u_i) \leq u_i \leq \bar{q}_i(u_i)$
To their payoff functions (economic, impulsion)	Economic impact without a feedback $s_i = const$	Economic impact with a feedback $s_i = s_i(u_i)$

Economic control mechanisms are implemented in the SPICE-model (1)–(2) by a choice of the values s_i by the Center. For administrative mechanisms it is supposed additionally that the Center can restrict feasible controls of the agents:

$$\tilde{q}_i \leq u_i \leq \bar{q}_i, i \in N. \tag{5}$$

Then the control mechanisms presented in Table 3 can be specified for the model (1)–(3), (5) as follows (Table 4).

3.1 Economic Mechanisms

Mechanisms Without a Feedback Suppose that $\forall i$ $s_i = const$. The first order conditions show that an internal system compatibility in this model is impossible: all players should be pure individualists or pure collectivists (Theorem 1).

Mechanisms with a Feedback Now assume that $s_i = s_i(u_i)$ or $s_i = s_i(u)$. The first order conditions show that an internal system compatibility in this model is possible only if

$$\frac{\partial s_i(u)}{\partial u_i} c(u) = [1 - s_i(u)] \frac{\partial c}{\partial u_i}, i \in N. \tag{6}$$

Two approaches are used for the further investigation: an empirical one and a theoretical one. The empirical approach analyzes the methods of distribution of the total income widely spread in practice. Consider, for example, a method of proportional distribution

$$s_i(u) = \begin{cases} \frac{u_i}{\sum_{j \in N} u_j}, & \exists m : u_m > 0 \\ 0, & otherwise. \end{cases}$$

Theorem 2 ([13]) *The mechanism of proportional distribution is system compatible only if the function of production of a social good is linear.*

Theorem 3 ([13]) *A mechanism of distribution can be system compatible if the function of production of a social good is symmetrical by u_{-i}.*

Corollary 1 *The mechanism of uniform distribution*

$$s_i(u) = \begin{cases} \frac{1}{|j:u_j=u_j^{\max}|}, & u_i = u_i^{\max}, \\ 0, & otherwise, \end{cases} \quad , i = 1, \dots, n,$$

is not system compatible.

Other economic control mechanisms are also possible. For example, the mechanism

$$s_i(u) = \begin{cases} \frac{1}{|j:u_j=r_j|}, & u_i = r_i, \\ 0, & otherwise, \end{cases} \tag{7}$$

allocates the social good only between pure collectivists. Notice that in this case all players have only two rational strategies: $\forall i : U_i = \{0, r_i\}$, so that mechanism (7) reduces a SPICE-model of the general type to the SPICE-model with binary sets of strategies.

Now let us formulate the problem of control mechanisms design in a general form. Assume that the Center maximizing the function of social welfare (3) reports to all players with payoff functions (1) a control mechanism

$$s_i(u) = \begin{cases} \frac{1}{|j:u_j=u_j^{\max}|}, & u_i = u_i^{\max}, \\ 0, & otherwise. \end{cases} \quad , i = 1, \dots, n, \tag{8}$$

Then the payers' payoffs are equal to

$$g_i(u) = \begin{cases} p_i(r_i - u_i^{\max}) + \frac{c(u_i^{\max}, u_{-i})}{|j:u_j=u_j^{\max}|}, & u_i = u_i^{\max}, \\ p_i(r_i - u_i), & otherwise. \end{cases}$$

It is evident that $U_i = \{0, u_i^{\max}\}$ because if $u_i > 0$, $u_i \neq u_i^{\max}$ then $g_i(u) = p_i(r_i - u_i) < p_i(r_i)$. Therefore, the mechanism (8) also reduces a SPICE-model of the general type to the SPICE-model with binary sets of strategies.

The problem is that player i in the moment of his decision does not know u_{-i} and the set $\{j : u_j = u_j^{\max}\}$, respectively. So, it is not simple to estimate an efficiency of the mechanism (8) (to compare the payoffs) in the general case. One can assert that the best response of player i to the mechanism (8) has the form

$$
u_i^{opt} = \begin{cases} u_i^{\max}, \ \forall u_{-i} \in U_{-i} & p_i(r_i) < p_i(r_i - u_i^{\max}) + \dfrac{c(u_i^{\max}, u_{-i})}{|j:u_j = u_j^{\max}|}, \\ 0, \ \ \forall u_{-i} \in U_{-i} & p_i(r_i) \geq p_i(r_i - u_i^{\max}) + \dfrac{c(u_i^{\max}, u_{-i})}{|j:u_j = u_j^{\max}|}, \end{cases} \tag{9}
$$

i.e. one of the feasible strategies dominates the other and is therefore the dominant one. But a question about the best response is open if for different u_{-i} the signs of inequalities are different (i.e. both strategies are non-dominated).

Denote $K = |\{j : u_j = u_j^{\max}\}|$ and consider the mechanism

$$
s_i(u) = \begin{cases} \frac{1}{K}, \ u_i = u_i^{\max}, \\ 0, \ otherwise, \end{cases}, i = 1, \ldots, n.
$$

The main trouble is that the value K is unknown because it is impossible to calculate the number of players who agree with the Center's requirements ($u_i = u_i^{\max}$) in advance. The value of K can be estimated approximately given that for the player who agrees with the Center it is profitable that the number of the same players be as small as possible.

Step 0. Denote $L = \{i : u_i = u_i^{\max}\}, K = |L|$. Numerate the players by decreasing of $p_i(r_i)$. The set L does not include the players for whom $p_i(r_i) > p_i(r_i - u_i^{\max}) + \frac{1}{n}c(u_1^{\max} + u_2^{\max} + \ldots + u_n^{\max})$. Let the first approximation of the value K be $K_0 = n$.

Step 1. Denote $\bar{u}^{\max 1} = (u_1^{\max}, u_2^{\max}, \ldots, u_n^{\max})$. Denote L_1 the set of players for whom last inequality is wrong, $K_1 = |L_1|$.

Step 2. Then find $\bar{u}^{\max 2} = (\underbrace{0, 0, \ldots, 0}_{n-K_1}, u_{n-K_1+1}^{\max 2}, \ldots, u_n^{\max 2})$ as a vector maximizing the value of the function $g_0(0, 0, \ldots, 0, u_{n-K_1+1}, \ldots, u_n)$. Form a set L_2 of those agents from L_1 for whom the inequality $p_i(r_i) > p_i(r_i - u_i^{\max 2}) + \frac{1}{K_1}c(u_{n-K_1+1}^{\max 2} + u_{n-K_1+2}^{\max 2} + \ldots + u_n^{\max 2})$ is wrong, $K_2 = |L_2|$.

Step 3. Then find $\bar{u}^{\max 3} = (\underbrace{0, 0, \ldots, 0}_{n-K_2}, u_{n-K_2+1}^{\max 3}, \ldots, u_n^{\max 3})$ as a vector maximizing the value of the function $g_0(0, 0, \ldots, 0, u_{n-K_2+1}, u_{n-K_2+2}, \ldots, u_n)$, form a set L_3 of those agents from L_2 for whom the inequality $p_i(r_i) > p_i(r_i - u_i^{\max 3}) + \frac{1}{K_2}c(u_{n-K_2+1}^{\max 3} + u_{n-K_2+2}^{\max 3} + \ldots + u_n^{\max 3})$ is wrong.

Step i. And so on for L_i ($i = 4, 5, \ldots$) until $K_i = K_{i-1}$. In this case the algorithm stops and the last found value $\bar{u}^{\max i}$ is the optimal solution.

Notice that in the case of a linear function of the production of a social good each agent knows his critical value K^* such that for $K \leq K^*$ it is profitable for him to agree with the Center. Then the problem is solved in one step.

Illustrate the work of this algorithm on numerical examples.

Example 1 Suppose that there are $n = 10$ agents with linear functions of private income, and the production of a social good is described by a linear function, too.

The coefficients are equal, respectively, to $p_1 = 20$, $p_2 = 18$, $p_3 = 16$, $p_4 = 14$, $p_5 = 12$, $p_6 = 10$, $p_7 = 8$, $p_8 = 6$, $p_9 = 4$, $p_{10} = 2$, $c = 15$, and all $r_i = 2$.

Then the social welfare function has the form

$$g_0 = 20(2 - u_1) + 18(2 - u_2) + 16(2 - u_3) + 14(2 - u_4) + 12(2 - u_5) +$$

$$+10(2 - u_6) + 8(2 - u_7) + 6(2 - u_8) + 4(2 - u_9) + 2(2 - u_{10}) +$$

$$+15(u_1 + u_2 + u_3 + u_4 + u_5 + u_6 + u_7 + u_8 + u_9 + u_{10}),$$

and the payoff functions of all agents are

$$g_i = p_i(2 - u_i) + \frac{15}{K}(u_1 + u_2 + u_3 + u_4 + u_5 + u_6 + u_7 + u_8 + u_9 + u_{10}).$$

Maximization of the social welfare function gives

$$u_i^{\max} = \begin{cases} 2, & p_i < c, \\ 0, & p_i > c. \end{cases}$$

In this example it means that

$$u_1^{\max} = u_2^{\max} = u_3^{\max} = 0; \; u_4^{\max} = u_5^{\max} = u_6^{\max} = u_7^{\max} = u_8^{\max} = u_9^{\max} = u_{10}^{\max} = 2.$$

For each agent $u_i^{NE} = \begin{cases} 2, & K < \frac{c}{p_i}, \\ 0, & K > \frac{c}{p_i}. \end{cases}$

It can be seen that the first, second, and third players agree with the strategy $u_i = 0$ if K is greater than 15/20, 15/18, and 15/16, respectively. Given K is the integer, these players agree with the Center's strategy for any K. The players four, five, six, and seven agree to assign all their resources $r_i = 2$ for the production of a social good if K = 1 (i.e., if each of them will receive the total social good). The player eight agrees to assign all resources for the production of a social good if $K < \frac{15}{6}$, or $K \leq 2$ (he is ready to share the social good with only one other player). The condition for the player nine is $K < \frac{15}{4} < 4$, or he is ready to allocate the social good between three players. At last, for the tenth player $K < \frac{15}{2} < 8$, or he is ready to have six partners. Now from four numbers K = 1, K = 2, K = 3, K = 7 we choose those which correspond to the number of players pretending for the social good. When K = 1 this number is equal to ten that is impossible. When K = 7 this number is equal to four, K = 2—six, K = 3—five. Thus, the only acceptable variant is K = 7. In this case

$$g_0(u^{NE}) = 20 \cdot 2 + 18 \cdot 2 + 16 \cdot 2 + 14 \cdot 2 + 12 \cdot 2 + 10 \cdot 2 + 8 \cdot 2 + 6 \cdot 2 + 4 \cdot 2 +$$

$$+15(0 + 0 + 0 + 0 + 0 + 0 + 0 + 0 + 0 + 2) = 246 < g_0(u^{\max}) = 318.$$

Example 2 Suppose that there are $n = 5$ agents, all functions are square roots. The coefficients are equal to $p_1 = 14.7$, $p_2 = 14.5$, $p_3 = 14$, $p_4 = 13.5$, $p_5 = 13$, $c = 10.5$, and all $r_i = 2$. Then the function of social welfare has the form:

$$g_0 = 14.7\sqrt{2 - u_1} + 14.5\sqrt{2 - u_2} + 14\sqrt{2 - u_3} + 13.5\sqrt{2 - u_4} + 13\sqrt{2 - u_5} +$$
$$+ 10.5\sqrt{u_1 + u_2 + u_3 + u_4 + u_5}$$

and the agents' payoff functions are

$$g_i = p_i\sqrt{2 - u_i} + \frac{10.5}{K}\sqrt{u_1 + u_2 + u_3 + u_4 + u_5}.$$

Estimate the value K. Start from the maximal value $K_0 = 5$. Calculate $u_i^{\max}$: $u_1^{\max} = 0.0063$, $u_2^{\max} = 0.0601$, $u_3^{\max} = 0.1916$, $u_4^{\max} = 0.3185$, $u_5^{\max} = 0.4407$, and $g_0^{\max} = 104.1076$.

If the agents agree to assign for the production of a social good the number of resources that is optimal to the center, then they share the produced social good equally, otherwise the agent does not participate in the consumption of the social good. So, each agent compares his payoffs in these two variants:

$$g_1(u_1^{\max}, u_{-1}^{\max}) = 22.8744 > 20.7889 = g_1(0, u_{-1}^{\max}),$$
$$g_2(u_2^{\max}, u_{-2}^{\max}) = 22.3134 > 20.5061 = g_2(0, u_{-2}^{\max}),$$
$$g_3(u_3^{\max}, u_{-3}^{\max}) = 20.9447 > 19.7990 = g_3(0, u_{-3}^{\max}),$$
$$g_4(u_4^{\max}, u_{-4}^{\max}) = 19.6239 > 19.0919 = g_4(0, u_{-4}^{\max}),$$
$$g_5(u_5^{\max}, u_{-5}^{\max}) = 18.3512 < 18.3848 = g_5(0, u_{-5}^{\max}).$$

So, only the fifth player refused to collaborate with the Center, and the social welfare function becomes equal to $g_0(u_1^{\max}, u_2^{\max}, u_3^{\max}, u_4^{\max}, 0) = 103.6415$.

The next approximation of the value K is $K_1 = 4$. In this case the following strategies of the agents are optimal for the Center:

$$u_1^{\max} = 0.1104, u_2^{\max} = 0.1614, u_3^{\max} = 0.2860, u_4^{\max} = 0.4063, u_5^{\max} = 0$$

and the agents' payoffs are

$$g_1(u_1^{\max 2}, u_{-1}^{\max 2}) = 22.7844 > 20.7889 = g_1(0, u_{-1}^{\max 2}),$$
$$g_2(u_2^{\max 2}, u_{-2}^{\max 2}) = 22.2387 > 20.5061 = g_2(0, u_{-2}^{\max 2}),$$
$$g_3(u_3^{\max 2}, u_{-3}^{\max 2}) = 20.9062 > 19.7990 = g_3(0, u_{-3}^{\max 2}),$$
$$g_4(u_4^{\max 2}, u_{-4}^{\max 2}) = 19.6201 > 19.0919 = g_4(0, u_{-4}^{\max 2}),$$
$$g_5(u_5^{\max 2}, u_{-5}^{\max 2}) = 18.3848 = 18.3848 = g_5(0, u_{-5}^{\max 2}).$$

Here the agents from one to four agree with the Center, and her payoff is
$g_0(u_1^{\max 2}, u_2^{\max 2}, u_3^{\max 2}, u_4^{\max 2}, 0) = 103.9343$. Thus, the final estimate is $K = 4$.

The theoretical approach is based on Germeier's theorem [15].

Theorem 4 ([13]) *If functions of the private payoff and the production of a social good are powers with an exponent less or equal than one, then the economic mechanism is system compatible.*

3.2 Administrative Mechanisms

Now suppose that the Center can constrain the sets of feasible controls of the agents. Consider only a case of the Stackelberg games without a feedback ($\Gamma 1$) because the case of feedback (inverse Stackelberg games, $\Gamma 2$) is more applicable for the formalization of corruption which is considered later. Then the model (1)–(3) takes the form

$$g_i(\tilde{q}_i, \bar{q}_i, u) = p_i(r_i - u_i) + s_i c(u) \to \max, \tilde{q}_i \le u_i \le \bar{q}_i, s_i \in [0, 1]; \quad (10)$$

$$g_0(\tilde{q}, \bar{q}, u) = \sum_{j \in N} p_j(r_j - u_j) + c(u) \to \max, 0 \le \tilde{q}_i \le \bar{q}_i \le r_i, i \in N. (11)$$

It is evident that if the possibilities of the Center are unrestricted then there is a trivial solution $\tilde{q}_i = \bar{q}_i = u_i^{\max}, i \in N$. Therefore an adequate setup of the problem of system compatible control mechanisms design requires a consideration of the Center's cost. Then (11) takes the form

$$g_0(\tilde{q}, \bar{q}, u) = \sum_{j \in N} p_j(r_j - u_j) + c(u) - C(\tilde{q}, \bar{q}) \to \max, 0 \le \tilde{q}_i \le \bar{q}_i \le r_i, i \in N,$$

where $C(\tilde{q}, \bar{q})$ is the continuously differentiable and convex by all arguments Center's administrative control cost function.

Definition 3 An administrative mechanism $\tilde{q}^{\max}(\bar{q}^{\max})$ is weakly compatible if $u = \tilde{q}^{\max} \in NE(\tilde{q}^{\max})$ and $g_0(\tilde{q}, \bar{q}, \tilde{q}^{\max}) = \max_{\tilde{q}^{\max} \le u_i \le r_i} g_0(\tilde{q}, r, u)$ (respectively, $u = \bar{q}^{\max} \in NE(\bar{q}^{\max})$ $g_0(\tilde{q}, \bar{q}, \bar{q}^{\max}) = \max_{0 \le u_i \le \bar{q}^{\max}} g_0(0, \bar{q}, u)$).

Theorem 5 *For an administrative mechanism in the model (10)–(11)* $\tilde{q}_i^{\max} \le u_i^{\max} \le \bar{q}_i^{\max}$.

Proof To find $u_i^{\max}$ it is necessary to solve the system of equations $-p_i'(r_i - u_i) + c'(u) = 0$, to find $\tilde{q}_i^{\max}$—the system $-p_i'(r_i - \tilde{q}_i) + c'(\tilde{q}) = C'_{\tilde{q}}(\tilde{q}, \bar{q}) > 0$, and to find $\bar{q}_i^{\max}$—the system of equations $-p_i'(r_i - \bar{q}_i) + c'(\bar{q}) = C'_{\bar{q}}(\tilde{q}, \bar{q}) < 0$. The same function in the left-hand part in all three cases decreases, and the values in the right-hand part are strictly ordered. Therefore, $\tilde{q}_i^{\max} \le u_i^{\max} \le \bar{q}_i^{\max}$. $\square$

Corollary 2 *It is senseless for the Center to restrict the agents from above.*

In fact, Theorem 1 says that $u_i^{NE} < u_i^{\max}$, and with consideration of the Theorem 5, $u_i^{\max} < \bar{q}_i^{\max}$. Therefore, the right side of the inequality $\tilde{q}_i \leq u_i \leq \bar{q}_i$ is evident.

Thus, only models with restrictions from below are considered:

$$g_i(q_i, u) = p_i(r_i - u_i) + s_i c(u) \rightarrow \max, \, q_i \leq u_i \leq r_i, s_i \in [0, 1]; \quad (12)$$

$$g_0(q, u) = \sum_{j \in N} p_j(r_j - u_j) + c(u) - C(q) \rightarrow \max, 0 \leq q_i \leq r_i, i \in N. \quad (13)$$

As cost functions we use linear $C(q) = \sum_{i=1}^{N} \alpha_i q_i$, quadratic $C(q) = \sum_{i=1}^{N} \alpha_i q_i^2$, and hyperbolic $C(q) = \sum_{i=1}^{N} \frac{\alpha_i q_i}{r_i - q_i}$ functions. A case of absence of the costs

$$g_i(q_i, u) = p_i(r_i - u_i) + s_i c(u) \rightarrow \max, \, q_i \leq u_i \leq r_i, s_i \in [0, 1];$$

$$g_0(q, u) = \sum_{j \in N} p_j(r_j - u_j) + c(u) \rightarrow \max, 0 \leq q_i \leq r_i, i \in N$$

means that the Center maximizes the social welfare function which is already analyzed in Sect. 2.2. In that case the agent's optimal strategy without consideration of the condition $q_i \leq u_i \leq r_i$ is u_i^*, and the Center's optimal strategy is $q_i = u_i^{\max}$. It is proved in Theorem 1 that $u_i^{\max} \geq u^*$; therefore, the agent's optimal strategy with consideration of the condition $q_i \leq u_i \leq r_i$ is $u_i = q_i = u_i^{\max}$ that gives the system compatibility.

Theorem 6 *An administrative mechanism in the model (12)–(13) is weakly compatible if for any $i \in N$ one of the following conditions is satisfied:*

$$\underset{u_i \in R}{Arg\max} \left[p_i(r_i - u_i) + s_i c \left(\sum_{i=1}^{n} u_i \right) \right] \leq$$

$$\underset{q_i \in R}{Arg\max} \left[\sum_{i=1}^{n} p_i(r_i - q_i) + c \left(\sum_{i=1}^{n} q_i \right) - C(q) \right]$$

or

$$\underset{u_i \in R}{Arg\max} \left[p_i(r_i - u_i) + s_i c \left(\sum_{i=1}^{n} u_i \right) \right] > r_i,$$

$$\underset{q_i \in R}{Arg\max} \left[\sum_{i=1}^{n} p_i(r_i - q_i) + c \left(\sum_{i=1}^{n} q_i \right) - C(q) \right] > r_i.$$

Proof Find the Nash equilibrium. The best response of an agent to the Center's strategy is

$$u_i^{NE} = \begin{cases} u_i^*, & q_i < u_i^* < r_i, \\ q_i, & q_i > u_i^*, \\ r_i, & u_i^* > r_i. \end{cases} \quad u_i^* = Arg\max_{u_i \in R}\left[p_i(r_i - u_i) + s_i c \left(\sum_{i=1}^n u_i\right)\right].$$

From the Center's point of view the optimal strategy of the agent is

$$u_i^{max} = \begin{cases} u_i^{**}, & q_i < u_i^{**} < r_i, \\ q_i, & q_i > u_i^{**}, \\ r_i, & u_i^{**} > r_i. \end{cases}$$

$$u_i^{**} = Arg\max_{q_i \in R}\left[\sum_{i=1}^n p_i(r_i - u_i) + c\left(\sum_{i=1}^n u_i\right) - C(q)\right].$$

There are unique values u_i^* and u_i^{**} due to negativity of the second derivatives of the functions g_i and g_0, and $u_i^{max} \geq u_i^{NE}$ (Theorem 1). Three cases are possible:

1. $u_i = r_i \Rightarrow q_i = 0$ and g_0 decreases by q_i;
2. $q_i < u_i < r_i \Rightarrow q_i = 0$;
3. $u_i = q_i \Rightarrow$ the optimal q_i is found as a solution of the problem

$$q_i^* = Arg\max_{q_i \in R}\left[\sum_{i=1}^n p_i(r_i - q_i) + c\left(\sum_{i=1}^n q_i\right) - C(q)\right],$$

and finally

$$q_i^* = \begin{cases} q_i^*, & 0 < q_i^* < r_i, \\ 0, & 0 > q_i^*, \\ r_i, & q_i^* > r_i. \end{cases}$$

□

4 Corruption in SPICE-Models

An author's approach to the modeling of corruption in resource allocation in the hierarchical control systems is proposed in [14]. It is worthwhile to distinguish administrative and economic corruption. A hierarchical control system with corruption includes three levels: principal, supervisor, and agents (Fig. 4).

The principal is not corrupted and controls the agents' activity by an influence to their sets of feasible strategies or payoff functions. But in fact the principal delegates his functions to the supervisor who can weaken in exchange for a bribe from the agents the administrative requirements (administrative corruption) or

Fig. 4 A hierarchical control system "principal-supervisor-agents"

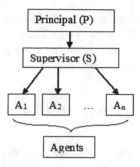

economic requirements (economic corruption). In real organizations there are many supervisors; we consider only one for simplicity.

4.1 Economic Corruption in SPICE-Models

Suppose that without corruption a social good $c(u)$ in the model (1)–(3) is distributed between the principal, the supervisor, and the agents in the proportion $t^0, r^0, \sum_{j=1}^{n} s_j^0$, where $t^0 + r^0 + \sum_{j=1}^{n} s_j^0 = 1$. In the case of economic corruption the supervisor increases a share of the agents in a social good at the expense of the principal:

$$t = t^0 - \sum_{j=1}^{n} \delta_j, r = r^0 + \sum_{j=1}^{n} b_j \delta_j, s_i = s_i^0 + (1-b_i)\delta_i, i \in N, \qquad (14)$$

and for the new shares of distribution (14) the conditions $t + r + \sum_{j=1}^{n} s_j = 1$, $t^0 - \sum_{j=1}^{n} \delta_j > 0, r^0 + \sum_{j=1}^{n} b_j \delta_j < 1, s_i = s_i^0 + (1-b_i)\delta_i < 1, i \in N$ should be satisfied. Here δ_i—an increase of the share of a social good for the i-th agent, b_i— a share of "kickback" of δ_i to the supervisor from the i-th agent. Then the model (1)–(3) takes the form

$$g_S(b, \delta, u) = [r^0 + \sum_{j=1}^{n} b_j \delta_j]c(u) \to \max, 0 \le \delta_i \le 1; \qquad (15)$$

$$g_i(b_i, \delta_i, u) = p_i(r_i - u_i) + [s_i^0 + (1-b_i)\delta_i]c(u) \to \max_{0 \le b_i \le 1, 0 \le u_i \le r_i}, i = 1, \ldots, n, \quad (16)$$

where g_S, g_i are payoff functions of the supervisor and the i-th agent, respectively.

The model (15)–(16) can be analyzed by two approaches: descriptive and normative. In the case of descriptive approach the form of a bribery function $\delta_i(b_i)$ is supposed to be known, and the agents play a game in normal form where the strategies are pairs of control variables (b_i, u_i). In the normative approach the function $\delta_i(b_i)$ is built as the optimal strategy of the supervisor in a game of the type $\Gamma 2$ with the agents.

Let us use the descriptive approach with linear and power functions $\delta_i(b_i)$ which are widely spread in the literature. Consider the model in general form:

$$g_i(b_i, u) = p_i(r_i - u_i) + [s_i^0 + (1 - b_i)\delta_i(b_i)]c(\sum_{j=1}^{n} u_j), i = 1, \ldots, n.$$

The first order conditions determine the optimal value b_i as the solution of the equation $-\delta_i(b_i) + (1 - b_i)\delta_i'(b_i) = 0$. Note that the optimal value does not depend on u_i, and the conditions $t^0 - \sum_{j=1}^{n} \delta_j > 0$, $r^0 + \sum_{j=1}^{n} b_j \delta_j < 1$, $s_i = s_i^0 + (1 - b_i)\delta_i < 1$, $i \in N$ should be satisfied, or the optimal solution is situated on the boundary of the feasible domain.

Let $u_i \in Arg \max \left[p_i(r_i - u_i) + [s_i^0 + (1 - b_i^*)\delta_i(b_i^*)]c(\sum_{j=1}^{n} u_j) \right]$.

Thus, the system compatibility is possible only when $u_i = 0$ or $u_i = r_i$, otherwise $s_i = 1$.

In the normative approach define the principal's payoff function

$$g_P(u) = \sum_{j=1}^{n} p_j(r_j - u_j) + c(u) \to \max \tag{17}$$

and consider the model (16), (17). The condition of system compatibility takes the form $u^* = u^{\max}$, where $u = (u_1, \ldots, u_n)$, u^* is a Nash equilibrium in the game (15), $g_P(u^{\max}) = \max_u g_P(u)$.

The following method of the model investigation based on Germeier's theorem [15] is proposed. First, the values u^*, $u_{\max}$ and the values s_i^0 such that u^*, $u_{\max}$ are as close as possible are calculated. Second, a solution of the game $\Gamma 2$ between the supervisor and the agent is built: $\delta_i^P = 0$; $E_i = \{b_i = 0\}$; $L_i = p_i(r_i - u_i^*) + s_i c(\sum_{i \in N} u_i^*)$—the maximal agent's payoff in the case of punishment, $u_i^* \in Arg \max_{u_i} \left[p_i(r_i - u_i) + s_i c(\sum_{j=1}^{n} u_j) \right]$.

$$K_2 = \max_{\delta_i} \min_{b_i \in E_i} \left[\left(r + \sum_{j=1}^{n} b_j \delta_j \right) c(u) \right] = \max_{\delta_i}[rc(u)] = rc(u^*)$$

$$K_1 = \max_{\delta_i} \max_{u_i} \max_{b_i} \left[\left(r + \sum_{j=1}^{n} b_j \delta_j \right) c(u) \right].$$

with constraint $p_i(r_i - u_i) + (s_i + (1 - b_i)\delta_i)c(u) > p_i(r_i - u_i^*) + s_i c(u^*)$

To determine K_1 it is necessary to solve a maximization problem with two variables: b_i and δ_i. The above constraint gives the dependence of b_i on δ_i:

$$b_i = 1 - \frac{p_i(r_i - u_i^*) - p_i(r_i - u_i) + s_i \left[c(\sum_{i \in N} u_i^*) - c(\sum_{i \in N} u_i) \right]}{\delta_i c(\sum_{i \in N} u_i)} - \epsilon_i.$$

Substitution of this value into $\left(r + \sum_{j=1}^n b_j \right) c(u) + \sum_{j=1}^n p_j(r_j - u_j)$ implies the maximization problem with one variable δ_i with constraint $\sum_{j=1}^n \delta_j < 1 - r_0 - \sum_{j=1}^n s_j^0$.

Example 3 Suppose that there are three agents, and all model functions are linear with coefficients $p_1 = 4$, $p_2 = 6$, $p_3 = 15$, $c = 14$. Let $r_i = 2$, $r = 0.02$. Then a social good production function takes the form:

$$g_0 = 4(2 - u_1) + 6(2 - u_2) + 15(2 - u_3) + 14(u_1 + u_2 + u_3).$$

First, the principal maximizes the social good production function that implies $u_1^{max} = 2$, $u_2^{max} = 2$, $u_3^{max} = 0$. The respective agents agree with the values if $s_1 > \frac{2}{7}$, $s_2 > \frac{3}{7}$, any s_3. Let $s_1 = 0.3$, $s_2 = 0.43$, $s_3 = 0$, $r = 0.02$. Due to $s_1 + s_2 < 1$ the system compatibility is possible. Therefore, $u_1^* = 2$, $u_2^* = 2$, $u_3^* = 0$, and the total payoff is equal to $14(2 + 2) = 56$, the principal's payoff $g_0 = 15 \cdot 2 + 14(2 + 2) = 86$, the agents' payoffs $g_1 = 0.3 \cdot 14(2 + 2) = 16.8$, $g_2 = 0.43 \cdot 14(2 + 2) = 24.08$, $g_3 = 15 \cdot 2 = 30$.

Now the supervisor solves her problem. She founds $K_2 = 0.02 \cdot 56 = 1.12$. Then

$$K_1 = \max_{\delta_i} \max_{u_i} \max_{b_i} [(0.02 + b_1\delta_1 + b_2\delta_2 + b_3\delta_3) \cdot 14 \cdot (u_1 + u_2 + u_3)]$$

with constraints

$$(0.03 + (1 - b_1)\delta_1) \cdot 14 \cdot (u_1 + u_2 + u_3) + 4(2 - u_1) > 16.8,$$
$$(0.43 + (1 - b_2)\delta_2) \cdot 14 \cdot (u_1 + u_2 + u_3) + 6(2 - u_2) > 24.08,$$
$$(1 - b_3)\delta_3 \cdot 14 \cdot (u_1 + u_2 + u_3) + 15(2 - u_3) > 20,$$

that implies

$$b_1 = 1 - \frac{0.3[56 - 14(u_1 + u_2 + u_3)] - 4(2 - u_1)}{\delta_1 \cdot 14(u_1 + u_2 + u_3)} - \epsilon,$$

$$b_2 = 1 - \frac{0.43[56 - 14(u_1 + u_2 + u_3)] - 6(2 - u_2)}{\delta_2 \cdot 14(u_1 + u_2 + u_3)} - \epsilon,$$

$$b_3 = 1 - \frac{30 - 15(2 - u_3)}{\delta_3 \cdot 14(u_1 + u_2 + u_3)} - \epsilon.$$

Substitution into the supervisor's payoff function gives

$$K_1 = \max_{\delta_i} \max_{u_i} [(0.02 + (1 - \epsilon)(\delta_1 + \delta_2 + \delta_3) \cdot 14 \cdot (u_1 + u_2 + u_3) -$$

$$- 0.3[56 - 14(u_1 + u_2 + u_3)] - 4(2 - u_1) - 0.43[56 - 14(u_1 + u_2 + u_3)]$$

$$- 6(2 - u_2) - 15u_3] = \max_{\delta_i} \max_{u_i} [(0.02 + (1 - \epsilon)(\delta_1 + \delta_2 + \delta_3) \cdot 14 \cdot (u_1 + u_2 + u_3) -$$

$$- 16.8 + 4.2(u_1 + u_1 + u_3) + 8 - 4u_1 - 24.08 + 6.02(u_1 + u_2 + u_3) + 12 - 6u_2 -$$

$$- 15u_3] = \max_{\delta_i} \max_{u_i} [(0.02 + (1 - \epsilon)(\delta_1 + \delta_2 + \delta_3) \cdot 14 \cdot (u_1 + u_2 + u_3) -$$

$$- 20.88 + 6.22u_1 + 4.22u_2 - 4.8u_3]$$

and the optimal values are: $\delta_1 + \delta_2 + \delta_3 = 1 - 0.2 - 0.3 - 0.43 = 0.25$, $u_1 = u_1^* = 2$, $u_2 = u_2^* = 2$, $u_3 = u_3^* = 0$. Follow the algorithm, set $\delta_1 = 0.125$, $\delta_2 = 0.125$, $\delta_3 = 0$. Then $K_1 = \max_{\delta_i} \max_{u_i} [(0.02 + (1 - \epsilon) \cdot 0.125) \cdot 14 \cdot 4 - 20.88 + 6.22 \cdot 2 + 4.22 \cdot 2] = 7(1 - \epsilon) + 1.12 > K_2$, the agents choose the controls $b_1 = b_2 = b_3 = 1 - \epsilon$, and their payoff functions take the values $g_1 = (0.3 + \epsilon \cdot 0.125) \cdot 56 = 16.8 + 7\epsilon > 16.8$, $g_2 = (0.43 + \epsilon \cdot 0.125) \cdot 56 = 24.08 + 7\epsilon > 24.08$, $g_3 = 15 \cdot 2 = 30$. The principal's payoff is equal to $g_0 = 15 \cdot 2 + 14(2 + 2) = 86$.

4.2 Administrative Corruption in SPICE-Models

The analysis of the administrative corruption is based on the SPICE-model in the form

$$g_S(\epsilon, b, u) = rc(u) + \sum_{j=1}^{n} b_j p_j (r_j - u_j) \to \max, 0 \le \epsilon_i \le q_i; \tag{18}$$

$$g_i(\epsilon_i, b_i, u) = (1 - b_i) p_i (r_i - u_i) + s_i c(u) \to \max_{0 \le b_i \le 1, q_i - \epsilon_i \le u_i \le r_i}, i = 1, \ldots, n. \tag{19}$$

The principal bounds a choice of the agent u_i from below by the value q_i but the supervisor weakens the bound on the value ϵ_i in exchange to the "kickback" $b_i p_i (r_i - u_i)$ from the agent's payoff.

As in the case of economic corruption, a bribery function $\epsilon_i(b_i)$ can be defined in two ways. The descriptive approach means that it is given, while in the normative approach it is calculated as an optimal guaranteeing strategy in a game $\Gamma 2$ of the supervisor with the agents. It is natural to suppose that $\epsilon_i(b_i)$ increases monotonically on the segment $[0,1]$, $\epsilon_i(0) = 0$, $\epsilon_i(1) = q_i$.

The value q_i in the model (18)–(19) is determined by the principal. Then the supervisor defines only a form of the function $\epsilon_i(b_i)$. Except the variable b_i, this function may contain additional parameters which depend or do not depend on the supervisor's actions. If a parameter is controllable by the supervisor, she can choose an optimal value of it. An agent solves the following optimization problem:

$$u_i^*(b_i) \in Arg \max \left[(1 - b_i) p_i (r_i - u_i) + s_i c(\sum_{j=1}^{n} u_j) \right]$$

or with the consideration of constraints

$$u_i = \begin{cases} u_i^*(b_i), & q_i - \epsilon_i(b_i) < u_i^*(b_i) < r_i, \\ q_i - \epsilon_i(b_i), & q_i - \epsilon_i(b_i) > u_i^*(b_i), \\ r_i, & u_i^*(b_i) > r_i. \end{cases}$$

In the case of the second branch the agent has no need to diminish the bound q_i. In the case of the first branch it is necessary to find the value b_i such that $q_i - \epsilon_i(b_i) = u_i^*(b_i)$. A greater bribe is senseless because $u_i > q_i - \epsilon_i(b_i)$. If from the very beginning $u_i(b_i) > q_i$, then $b_i = 0$.

In the normative approach to the model

$$g_P(\epsilon, b, u) = c(u) + \sum_{j=1}^{n} p_j (r_j - u_j) \to \max, 0 \le q_i \le r_i;$$

$$g_S(\epsilon, b, u) = rc(u) + \sum_{j=1}^{n} b_j p_j (r_j - u_j) \to \max, 0 \le \epsilon_i(b_i) \le q_i;$$

$$g_i(\epsilon_i, b_i, u) = (1 - b_i) p_i (r_i - u_i) + s_i c(u) \to \max, 0 \le b_i \le 1, q_i - \epsilon_i(b_i) \le u_i \le r_i,$$

$$i = 1, \ldots, n$$

Germeier's theorem is applied [15] similar to the previous paragraph that gives

$$\epsilon_i^P = 0; \ E_i = \{b_i = 0\}; \ L_i = p_i (r_i - u_i^*) + s_i c(\sum_{i \in N} u_i^*)$$

$$K_2 = \max_{\epsilon_i} \min_{b_i \in E_i} \left[rc(u) + \sum_{j=1}^{n} b_j p_j (r_j - u_j) \right] = \max_{\epsilon_i} [rc(u)] = rc(u)$$

$$K_1 = \max_{\epsilon_i} \max_{u_i} \max_{b_i} \left[rc(u) + \sum_{j=1}^{n} b_j p_j (r_j - u_j) \right]$$

with constraint $(1 - b_i) p_i (r_i - u_i) + s_i c(u) > p_i (r_i - u_i^*) + s_i c(u^*)$.

To determine K_1 it is necessary to solve a maximization problem with two variables: b_i and δ_i. The above constraint gives the dependence of b_i on δ_i:

$$b_i = \frac{p_i (r_i - u_i) - p_i (r_i - u_i^*) + s_i \left[c(\sum_{i \in N} u_i) - c(\sum_{i \in N} u_i^*) \right]}{p_i (r_i - u_i)} - \epsilon_i.$$

Substitution of the value into the expression $rc(u) + \sum_{j=1}^{n} b_j p_j(r_j - u_j)$ gives a standard optimization problem. If necessary $(u_i^* < q_i)$, then its solution gives the optimal value ϵ_i, otherwise $(u_i^* > q_i)$ $\epsilon_i = 0$. Thus, an agent needs the supervisor's help only if $u_i^* < q_i$.

5 SPICE-Models in the Regional Development

Consider an application of the SPICE-models to the problems of the development of trans-frontier territories. There are two equal agents A_1 and A_2 with a common frontier. Each of them has a number of resources r_1 and r_2, respectively. The resources are assigned to their private economic activities and to the development of the trans-frontier territory (joint projects) in a ratio. Payoff functions of the agents consist of two summands: a private income and a share of the income from the joint project. The resources r_1 and r_2 may be fixed by a Center of the development of the trans-frontier territory who disposes a number of resources R. The Center can also determine the shares of the agents in the joint income or control a distribution of the resources by each of them.

The system is presented in Fig. 5. The payoff function of each agent has the form

$$g_i(r_i, s_i, u) = p_i(r_i - u_i) + s_i c(u_1 + u_2) \to \max, i = 1, 2,$$

where $p_i(r_i - u_i)$—a private income of the i-th agent (from the activity on his territory); $c(u_1, u_2)$—a common income from the joint projects (their activity on the trans-frontier territory); s_i—a share of the i-th agent in the common income.

For the purposes of control and coordination of the egoistic interests of the agents they organize a Center (or this can be done by a control agency of the higher level, for example, the federal state) and delegate her some administrative and/or economic power. Thus, the Center can use the following control mechanisms:

Fig. 5 Coordination of interests in the problem of control of trans-frontier territories

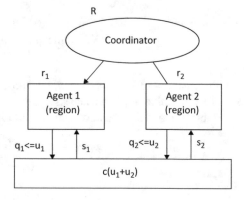

1. an administrative one when she fixes the lower bound q_i for the private use of resources ($q_i \leq u_i \leq r_i$) that incurs her control cost. In this case the Center's problem has the form

$$g_0(q, r, s, u) = g_1(r_1, s_1, u) + g_2(r_2, s_2, u) - C(q_1, q_2) \rightarrow \max$$

with one of the constraints:

$$0 \leq q_i \leq r_i; 0 \leq s_i \leq 1, s_1 + s_2 = 1; 0 \leq r_i \leq 1, r_1 + r_2 = R;$$

2. an economic one in two variants:

   (a) the Center fixes a share s_i of the agent in the common income;
   (b) she allocates a number of resources r_i to the each agent.

So, we have a hierarchical game of the Center with two agents (in fact, several agents are possible).

5.1 Administrative Control Mechanisms

The hierarchical game-theoretic model takes the form:

$$g_0(u) = g_1(u) + g_2(u) \rightarrow \max, \tag{20}$$

$$g_i(r_i, u) = p_i(r_i - u_i) + s_i c(u_1 + u_2) \rightarrow \max, \tag{21}$$

with constraints:

$$0 \leq q_i \leq r_i, q_i \leq u_i \leq r_i, i = 1, 2. \tag{22}$$

Notice that the Center's payoff function $g_0(u)$ does not depend explicitly on her strategy q but a choice of q restricts the sets of feasible strategies of the agents that can increase her payoff. It is assumed that the production functions $p_i(\cdot)$ and $c(\cdot)$ satisfy the following conditions (on the example of $p_i(\cdot)$):

1. $p_i(\cdot) \geq 0, p_i(\cdot) = 0$;
2. $p_i(\cdot)$ is continuous and increases monotonically, i.e. $p_i'(\cdot) > 0$;
3. $p_i(\cdot)$ is concave, i.e. $p_i''(\cdot) < 0$;
4. $p_i(\lambda \cdot) = \lambda^\alpha p_i(\cdot)$, where α is a coefficient of elasticity of the production process (assume for simplicity that $\alpha_1 = \alpha_2 = \alpha, 0 < \alpha \leq 1$ because the territories have close natural and economic characteristics).

These properties are widely used in the economic research and are represented by power functions with an exponent less or equal than one. Thus, four variants are convenient to analyze:

1. functions $p_i(\cdot)$, $c(\cdot)$, are linear;
2. functions $p_i(\cdot)$ are linear, and the function $c(\cdot)$ is a power one with an exponent strictly less than one;
3. the functions $p_i(\cdot)$ are power ones with an exponent strictly less than one, and the function $c(\cdot)$ is linear;
4. the functions $p_i(\cdot)$, $c(\cdot)$ are power ones with an exponent strictly less than one.

Theorem 7 *Suppose that in the model (20)–(22) the functions $p_i(\cdot)$, $c(\cdot)$ are increasing and concave, $p_i(0) = 0$, $c(0) = 0$, and all functions are not in the same time power ones with an exponent strictly less than one. Then Nash strategies of the agents coincide with the Pareto-optimal strategy of the Center: $u_i^{NE} = u_i^{\max}$.*

Proof

$$u_i^{NE} = \begin{cases} u_i^*, & 0 < u_i^* < r_i, \\ 0, & u_i^* < 0, \\ r_i, & u_i^* > r_i. \end{cases} \quad (23)$$

$$q_i = u_i^{\max} = \begin{cases} u_i^{**}, & 0 < u_i^{**} < r_i, \\ 0, & u_i^{**} < 0, \\ r_i, & u_i^{**} > r_i \end{cases} \quad (24)$$

where

$$u_i^* = \left(s_i c' \left(\sum_{i=1}^{n} u_i \right) - p_i'(r_i - u_i) \right)^{-1} (0),$$

$$u_i^{**} = \left(c' \left(\sum_{i=1}^{n} u_i \right) - p_i'(r_i - u_i) \right)^{-1} (0).$$

As far as $s_i \leq 1$ then $s_i c' \left(\sum_{i=1}^{n} u_i \right) \leq c' \left(\sum_{i=1}^{n} u_i \right)$, therefore, $s_i c' \left(\sum_{i=1}^{n} u_i \right) - p_i'(r_i - u_i) \leq c' \left(\sum_{i=1}^{n} u_i \right) - p_i'(r_i - u_i)$. Let $f(u_i) = s_i c' \left(\sum_{i=1}^{n} u_i \right) - p_i'(r_i - u_i)$, $g(u_i) = c' \left(\sum_{i=1}^{n} u_i \right) - p_i'(r_i - u_i)$. The functions $f(u_i)$ and $g(u_i)$ decrease, therefore, their inverse functions also decrease. Thus, the value of the mapping of the point 0 in the greater function $g(u_i)$ is not less of the respective value of the point 0 in the lesser function $f(u_i)$. Therefore, the Nash strategy of the agent i with consideration of the condition $q_i \leq u_i \leq r_i$ is $u_i = q_i = u_i^{\max}$.

5.2   Economic Control Mechanisms

In this case the model (20)–(21) is investigated with constraints

$$0 \leq s_i \leq 1, s_1 + s_2 = 1. \quad (25)$$

Denote for convenience $s_1 = s$, $s_2 = 1 - s$. If at least one of the functions $p_i(\cdot)$ or $c(x)$ is linear, then the following algorithm of investigation is proposed.

Stage 1 Found all Nash equilibria in the game of agents (21). If there are two agents, then not more than nine Nash outcomes are possible. The outcomes are separated from each other by the curves $\left(s_i c' \left(\sum_{i=1}^n u_i\right) - p_i'(r_i - u_i)\right)^{-1}(0) = 0$ and $\left(s_i c' \left(\sum_{i=1}^n u_i\right) - p_i'(r_i - u_i)\right)^{-1}(0) = r_i$. Each outcome is defined by the following two conditions:

1. one of the inequalities

$$\left(s_1 c' \left(\sum_{i=1}^n u_i\right) - p_1'(r_1 - u_1)\right)^{-1}(0) < 0,$$

$$0 < \left(s_1 c' \left(\sum_{i=1}^n u_i\right) - p_1'(r_1 - u_1)\right)^{-1}(0) < r_1$$

$$\left(s_1 c' \left(\sum_{i=1}^n u_i\right) - p_1'(r_1 - u_1)\right)^{-1}(0) > r_1;$$

holds,
2. and one of the inequalities

$$\left(s_2 c' \left(\sum_{i=1}^n u_i\right) - p_2'(r_2 - u_2)\right)^{-1}(0) < 0,$$

$$0 < \left(s_2 c' \left(\sum_{i=1}^n u_i\right) - p_2'(r_2 - u_2)\right)^{-1}(0) < r_2,$$

$$\left(s_2 c' \left(\sum_{i=1}^n u_i\right) - p_2'(r_2 - u_2)\right)^{-1}(0) > r_2.$$

holds.

Call these inequalities the conditions of feasibility of an outcome.

Stage 2 Solve the Center's optimization problem (20), (25). As in the previous stage, if there are two agents, then not more than nine Pareto-optimal outcomes are possible. The outcomes are separated from each other by the curves $\left(c' \left(\sum_{i=1}^n u_i\right) - p_i'(r_i - u_i)\right)^{-1}(0) = 0$ and $\left(c' \left(\sum_{i=1}^n u_i\right) - p_i'(r_i - u_i)\right)^{-1}(0) = r_i$. Each outcome is defined by the following two conditions are satisfied:

1. one of the inequalities

$$\left(c' \left(\sum_{i=1}^{n} u_i \right) - p_1'(r_1 - u_1) \right)^{-1} (0) < 0,$$

$$0 < \left(c' \left(\sum_{i=1}^{n} u_i \right) - p_1'(r_1 - u_1) \right)^{-1} (0) < r_1$$

$$\left(c' \left(\sum_{i=1}^{n} u_i \right) - p_1'(r_1 - u_1) \right)^{-1} (0) > r_1;$$

holds;

2. and one of the inequalities

$$\left(s_2 c' \left(\sum_{i=1}^{n} u_i \right) - p_2'(r_2 - u_2) \right)^{-1} (0) < 0,$$

$$0 < \left(s_2 c' \left(\sum_{i=1}^{n} u_i \right) - p_2'(r_2 - u_2) \right)^{-1} (0) < r_2,$$

$$\left(s_2 c' \left(\sum_{i=1}^{n} u_i \right) - p_2'(r_2 - u_2) \right)^{-1} (0) > r_2$$

holds.

Call 1 and 2 the conditions of optimality of an outcome. Consider that $u_i^{\max} > u_i^*$ we receive not more than 25 possible combinations of the outcomes (u_1^*, u_2^*) and $(u_1^{\max}, u_2^{\max})$.

Stage 3a In the specific case when the conditions of feasibility do not depend on the Center's control variables s, the problem is solved by an enumeration of all possible 25 variants.

Stage 3b If the conditions of feasibility depend on the Center's control variables, then the outcomes are ordered by decreasing of the Center's payoff function, and the conditions are checked for them sequentially. At least one outcome which satisfies the conditions exists, for example, when $s_1 = 0$, $s_2 = 1$.

If the model is analytically intractable, then the values of s are chosen by the following enumeration: for a given s from 0 till 1 with a given step the outcomes which satisfy the conditions of feasibility and optimality are chosen, and then it is determined

$$s^* = Arg \max_{s \in [0;1]} \max_{u_i \in D_i} g_0(u_1(s), u_2(s)),$$

where $D_i = \{u_i(s) | u_i(s) = Arg \max_{0 \le u_i(s) \le r_i} g_i(u)\}$.

Example 4 Suppose that there are two agents, the common interest is expressed by a linear function, and the private interests are described by power functions with an exponent less than one. The respective coefficients are $p_1 = 3$, $p_2 = 5$, $c = 4$, $\alpha = 1/2$, $r_1 = 2$, $r_2 = 2$.
   The agents' payoff functions are

$$g_1 = 3\sqrt{2 - u_1} + 4s(u_1 + u_2),$$

$$g_2 = 5\sqrt{2 - u_2} + 4(1 - s)(u_1 + u_2).$$

The Center's payoff function is

$$g_0 = 3\sqrt{2 - u_1} + 5\sqrt{2 - u_2} + 4(u_1 + u_2).$$

The agents' strategies which are optimal for the Center (Pareto-optimal):

$$u_1^{max} = 2 - \left(\frac{3}{8}\right)^2 = \frac{119}{64}, u_2^{max} = 2 - \left(\frac{5}{8}\right)^2 = \frac{103}{64}$$

and it is seen that the system compatibility is unavailable. It is only possible to maximize the price of anarchy.

The Center's payoff is $g_0 = 3\sqrt{\left(\frac{3}{8}\right)^2} + 5\sqrt{\left(\frac{5}{8}\right)^2} + 4 \cdot \frac{119+103}{64} = 18.125$.
   The Nash-strategies for the agents:

$$u_1^{NE} = \begin{cases} 2 - \frac{9}{64s^2}, & s > \frac{3}{8\sqrt{2}}, \\ 0, & s \le \frac{3}{8\sqrt{2}}. \end{cases} \quad u_2^{NE} = \begin{cases} 2 - \frac{25}{64(1-s)^2}, & s < 1 - \frac{5}{8\sqrt{2}}, \\ 0, & s \ge 1 - \frac{5}{8\sqrt{2}}. \end{cases}$$

Thus, in dependence of the value of s three outcomes are possible:

1. $\left(2 - \frac{9}{64s^2}; 2 - \frac{25}{64(1-s)^2}\right)$ if $\frac{3}{8\sqrt{2}} < s < 1 - \frac{5}{8\sqrt{2}}$, or (with precision up to 0.001) $0.265 < s < 0.558$.
2. $\left(2 - \frac{9}{64s^2}; 0\right)$ if $s \ge 1 - \frac{5}{8\sqrt{2}}$ or (with precision up to 0.001) $s \ge 0.558$.
3. $\left(0; 2 - \frac{25}{64(1-s)^2}\right)$ if $s \le \frac{3}{8\sqrt{2}}$ or (with precision up to 0.001) $s \le 0.265$.

   In the case 1 the Center's payoff takes the form

$$g_0 = 3\sqrt{\frac{9}{64s^2}} + 5\sqrt{\frac{25}{64(1-s)^2}} + 4\left(2 - \frac{9}{64s^2} + 2 - \frac{25}{64(1-s)^2}\right) =$$

$$= \frac{9}{8s} + \frac{25}{8(1-s)} + 16 - \frac{9}{16s^2} - \frac{25}{16(1-s)^2}.$$

This function attains its maximum when $\frac{\sqrt{3}}{\sqrt{3}+\sqrt{5}} \approx 0.436$ that satisfies the relation $\frac{3}{8\sqrt{2}} < s < 1 - \frac{5}{8\sqrt{2}}$. The Center's payoff is equal to $g_0 = 16.25$.

In the case 2 the Center's payoff takes the form

$$g_0 = 3\sqrt{\frac{9}{64s^2}} + 5\sqrt{2} + 4\left(2 - \frac{9}{64s^2}\right).$$

This function attains its maximum when $s = 1$ that satisfies the relation $s \geq 1 - \frac{5}{8\sqrt{2}}$. The Center's payoff is equal to $g_0 = \frac{9}{16} + 5\sqrt{2} + 8 \approx 15.633$.

In the case 3 the Center's payoff takes the form

$$g_0 = 3\sqrt{\frac{25}{64(1-s)^2}} + 3\sqrt{2} + 4\left(2 - \frac{25}{64(1-s)^2}\right).$$

This function attains its maximum when $s = 0$ that satisfies the relation $s \leq \frac{3}{8\sqrt{2}}$. The Center's payoff is equal to $g_0 = \frac{25}{16} + 3\sqrt{2} + 8 \approx 13.805$.

Thus, the maximal value of the Center's payoff function by s is attained when $\frac{\sqrt{3}}{\sqrt{3}+\sqrt{5}} \approx 0.436$ and is equal approximately to 16.25, and the price of anarchy is $PA = \frac{16.25}{18.125} \approx 0.897$. The respective strategies of the agents are: $u_1 = 2 - \frac{3(4+\sqrt{15})}{32} \approx 1.262$, $u_2 = 2 - \frac{5(4+\sqrt{15})}{32} \approx 0.770$, and their payoffs

$$g_1 = 3\sqrt{\frac{3(4+\sqrt{15})}{32}} + \frac{\sqrt{3}}{\sqrt{3}+\sqrt{5}}(12 - \sqrt{15}) \approx 6.120,$$

$$g_2 = 5\sqrt{\frac{5(4+\sqrt{15})}{32}} + \frac{\sqrt{5}}{\sqrt{3}+\sqrt{5}}(12 - \sqrt{15}) \approx 10.130.$$

5.3   Mechanisms of Resource Allocation

At last, in the context of the model (Fig. 5) the Center can give subsidies for some joint projects of the development of trans-frontier territories. Another example is an implementation of the projects of state-private partnership.

In this case the Center allocates to each agent a number of resources r_i. The whole received common income is divided among the agents. The model is the hierarchical game (20)–(21) with constraints

$$0 \leq r_i \leq 1, r_1 + r_2 = 1. \tag{26}$$

Denote $r_1 = r$, $r_2 = 1 - r$. The algorithm of investigation is similar to the one described in the previous paragraph. Consider a numerical example.

Example 5 Suppose that there are two agents, the common interest is expressed by a linear function, and the private interests are described by power functions with an exponent less than one. The respective coefficients are $p_1 = 3$, $p_2 = 5$, $c = 4$, $\alpha = 1/2$, $s_1 = 0.4$, $s_2 = 0.6$. The agents' payoff functions are

$$g_1 = 3\sqrt{r - u_1} + 1.6(u_1 + u_2),$$
$$g_2 = 5\sqrt{1 - r - u_2} + 2.4(u_1 + u_2).$$

The Center's payoff function is

$$g_0 = 3\sqrt{r - u_1} + 5\sqrt{1 - r - u_2} + 4(u_1 + u_2).$$

The agents' strategies which are optimal for the Center (Pareto-optimal):

$$u_1^{max} = \begin{cases} r - \frac{9}{64}, & r > \frac{9}{64}, \\ 0, & r \leq \frac{9}{64}. \end{cases} \quad u_2^{max} = \begin{cases} 1 - r - \frac{25}{64}, & r < \frac{39}{64}, \\ 0, & r \geq \frac{39}{64}. \end{cases}$$

Thus, three outcomes are possible:

1. $\left(r - \frac{9}{64}; 1 - r - \frac{25}{64}\right)$ if $\frac{9}{64} < r < \frac{39}{64}$;
2. $\left(r - \frac{9}{64}; 0\right)$ if $r \geq \frac{39}{64}$;
3. $\left(0; 1 - r - \frac{25}{64}\right)$ if $r \leq \frac{9}{64}$

and it is seen that the system compatibility is unavailable. It is only possible to maximize the price of anarchy. The Nash-strategies for the agents:

$$u_1^{NE} = \begin{cases} r - \frac{9}{10.24}, & r > \frac{9}{10.24}, \\ 0, & r \leq \frac{9}{10.24}. \end{cases} \quad u_2^{NE} = 0.$$

Thus, two outcomes are possible (in dependence on the value of r):

1. $\left(r - \frac{9}{10.24}; 0\right)$ if $r > \frac{9}{10.24}$;
2. $(0; 0)$ if $r \leq \frac{9}{10.24}$.

In the case 1 the Center's payoff takes the form

$$g_0 = 3\sqrt{\frac{9}{10.24}} + 5\sqrt{1-r} + 4\left(r - \frac{9}{10.24}\right) = 5\sqrt{1-r} + 4r - \frac{1}{12.8}.$$

This function attains its maximum when $r = \frac{39}{64} \approx 0.609$ that does not satisfy the relation $r > \frac{9}{10.24}$. In the case 2 the Center's payoff takes the form $g_0 = 3\sqrt{r} + 5\sqrt{1-r}$.

This function attains its maximum when $r = \frac{9}{34} \approx 0.265$ that satisfies the relation $r < \frac{9}{10.24}$. The Center's payoff is $g_0 = \sqrt{34} \approx 5.831$.

Thus, the maximal value of the Center's payoff function by r is attained when $r = \frac{9}{34}$ and is equal approximately to 5.831, and the maximal value of the Center's payoff when r is fixed (absence of control) is equal to 6.126; therefore, the price of anarchy is equal to $PA = \frac{5.831}{6.126} \approx 0.952$. The respective agents' strategies are: $u_1 = 0$, $u_2 = 0$, and their payoffs are

$$g_1 = 3\sqrt{\frac{9}{34}} = \frac{9}{\sqrt{34}},$$

$$g_2 = 5\sqrt{\frac{25}{34}} = \frac{25}{\sqrt{34}}.$$

6 Conclusion

The paper is dedicated to the problem of coordination of private and social interests in resource allocation. The previous results are presented together with the new ones.

In the considered SPICE-models all agents divide their resources between their private activities and the production of a common social good. Respectively, the payoff functions include both an income from the private activity and a share in the produced social good.

A function of social welfare is defined here as the sum of payoff functions of all agents. Then the key problem is system compatibility. It means that egoistic solutions of the agents (for example, their dominant or Nash strategies) in their totality maximize the social welfare. In other terms, the price of anarchy in the system should be as close to one as possible. It is proved that the system compatibility is attained only if the agents are divided into two sets: individualists, who assign all their resources to the private activities, and collectivists, who allocate all their resources to the production of a social good.

It is quite evident that the condition is a very strong commitment which does not hold in real economic and organizational systems as a rule. That is why special administrative and economic control mechanisms are necessary to provide the system compatibility. A specific agent (Center) is appointed as a social planner who solves the problem. In SPICE-models the Center can bound from below the sets of feasible strategies of the agents to restrict their egoism (administrative control mechanisms, compulsion) or fix the shares of the agents in the social good (economic control mechanisms, impulsion). Some results about the properties of the mechanisms are presented in the paper.

We consider corruption in hierarchical organizational systems as a feedback by the value of bribe (a bribery function) that leads to the inverse Stackelberg games. In the case of administrative corruption, the restrictions on the set of feasible strategies of an agent can be weakened in exchange for a bribe. In the case of economic corruption, a bribe permits to increase the share of an agent in a social good. In the analysis of corruption we use a descriptive approach when a bribery function is given from the experience, and a normative approach when the value of bribe is determined as a solution of an optimization problem or a game. The results about the connection of corruption with system compatibility are presented in the paper.

One of the most prospective domains of applications of the SPICE-models is the regional development, namely the problems of trans-frontier cooperation, state-private partnership, and others. Here it is also natural to introduce a Center for coordination of the agents. The economic, administrative, and resource allocation control mechanisms of such coordination are considered.

Acknowledgement The paper is supported by the Russian Science Foundation, project #17-19-01038.

References

1. Basar T., Olsder G.J.: Dynamic Noncooperative Game Theory. SIAM (1999)
2. Bergstrom, T., Blume C., Varian, H.: On the private provision of public goods. Journal of Public Economics. **29**, 25–49 (1986)
3. Bergstrom, T., Blume C., Varian, H.: Uniqueness of Nash Equilibrium in private provision of public goods: an improved proof. Journal of Public Economics. **49**, 391–392 (1992)
4. Bergstrom, T., Blume C., Varian, H.: When are Nash Equilibria independent of the distribution of agents' characteristics? Review of Economic Studies. **52**, 715–718 (1992)
5. Bernheim, B., Bagwell K.: Is everything neutral? Journal of Political Economy. **96**, 308–338 (1988)
6. Boadway, R., Pestiau P., Wildasin D.: Non-cooperative behavior and efficient provision of public goods. Public Finance. **44**, 1–7 (1989)
7. Boadway, R., Pestiau P., Wildasin D.: Tax-transfer policies and the voluntary provision of public goods. Journal of Public Economics. **39**, 157–176 (1989)
8. Burkov V.N., Lerner A.Ya.: Fair play in control of active systems. Differential games and related topics. Amsterdam, London: North-Holland Publishing Company, 164–168 (1971)
9. Christodoulou G., Sgouritza A., Tang B.: On the Efficiency of the Proportional Allocation Mechanism for Divisible Resources. M. Hoefer (Ed.): SAGT 2015, LNCS 9347, 165–177 (2015)
10. Germeier Yu.B., Vatel I.A.: Equilibrium situations in games with a hierarchical structure of the vector of criteria. Lecture Notes in Computer Science. **27**, 460–465 (1975)
11. Gorbaneva O.I., Ougolnitsky G.A.: Purpose and Non-Purpose Resource Use Models in Two-Level Control Systems. Advances in Systems Science and Applications. **13(4)**, 379–391 (2013)

12. Gorbaneva O.I., Ougolnitsky G.A.: System Compatibility, Price of Anarchy and Control Mechanisms in the Models of Concordance of Private and Public Interests. Advances in Systems Science and Applications. **15(1)**, 45–59 (2015)

13. Gorbaneva O.I., Ougolnitsky G.A.: Static Models of Coordination of Social and Private Interests in Resource Allocation. Automation and Remote Control. **78(10)**, 345–363 (2017)

14. Gorbaneva O.I., Ougolnitsky G.A., Usov A.B.: Modeling of Corruption in Hierarchical Organizations. N.Y.: Nova Science Publishers (2016)

15. Gorelik V.A., Kononenko A.F.: Game Theoretic Models of Decision Making in the Ecologic-Economic Systems. M. (1982) (in Russian)

16. International Handbook on the Economics of Corruption. Ed. by S. Rose-Ackerman. Edward Elgar (2006)

17. Kahana N., Klunover D.: Private provision of a public good with a time-allocation choice. **47**, 379–386 (2016)

18. Kemp M.: A note on the theory of international transfers. Economic Letters. **14, 2–3**, 259–262 (1984)

19. Kukushkin N.S.: A Condition for Existence of Nash Equilibrium in Games with Private and Public Objectives. Games and Economic Behavior. **7**, 177–192 (1994)

20. Mechanism Design and Management: Mathematical Methods for Smart Organizations. Ed. by Prof. D. Novikov. N.Y.: Nova Science Publishers (2013)

21. Novikov D.: Theory of Control in Organizations. N.Y.: Nova Science Publishers (2013)

22. Olsder G. J.: Phenomena in inverse Stackelberg games, part 1: Static problems. Journal of Optimization Theory and Applications. **143(3)**, 589–600 (2009)

23. Olsder G. J.: Phenomena in inverse Stackelberg games, part 2: Dynamic problems. Journal of Optimization Theory and Applications. **143(3)**, 601–618 (2009)

24. Ougolnitsky G.A.: Sustainable Management. N.Y.: Nova Science Publishers (2011)

25. Ricq Ch. Handbook of Transfrontier Co-operation, University of Geneva. Part III (2006)

26. Samuelson P.A.: The pure theory of public expenditure. Review of Economics and Statistics. **36**, 387–389 (1954)

27. Warr P.: The private provision of a public good is independent of the distribution of income. Economic Letters. **13**, 207–211 (1983)

Part III
Games on Graphs and Networks

A Multi-Stage Model of Searching for Two Mobile Objects on a Graph

Vasily V. Gusev

Abstract We are dealing with a search game where one searcher looks for two mobile objects on a graph. The searcher distributes his searching resource so as to maximize the probability of detecting at least one of the mobile objects. Each mobile object minimizes its own probability of being found. In this problem the Nash equilibrium, i.e. the optimal transition probabilities of the mobile objects and the optimal values of the searcher's resource, was found. The value of the game in a single-stage search game with non-exponential payoff functions was found.

Keywords Search theory · Game theory · Search model · Search on the graph · Dynamic search · Multi-stage game

1 Problem Statement and Notation

Numerous search models have been developed. In discrete-time models, the search will most often be run on a graph. We assume there is one searching player and two hiding players. In [5], on the contrary, there were several searchers and only one hider. The paper [2] considers a game of search for an immobile object on a cycle. In this article the graph is a tree and the hiders are mobile. The assumption in [1] is that hiders can turn around and move backwards. Here, we are dealing with mobile players who do not return to the nodes they have already been to. There are also studies [4] where the target is singular, but some noises make the searcher believe there are several mobile objects. Sometimes, special pathways are analyzed, e.g. Eulerian paths [3], to investigate the optimal strategies. Probability recalculation can yield Bayesian search models [6]. A more complete account of search-related literature can be found in [7].

V. V. Gusev (✉)
National Research University, Higher School of Economics, St. Petersburg, Russian Federation
e-mail: gusev@krc.karelia.ru

© Springer Nature Switzerland AG 2020
D. Yeung et al. (eds.), *Frontiers in Games and Dynamic Games*, Annals of the
International Society of Dynamic Games 16,
https://doi.org/10.1007/978-3-030-39789-0_5

Consider a multi-stage non-cooperative search game with three players (two mobile objects and a searcher) on a tree graph. The mobile objects are hiders numbered as first and second. Both mobile objects take start from the root vertex of the graph. We assume the searcher is watching the graph from above. At one stage each mobile object moves independently from one another from vertex i (mobile object's position) to vertex j with some probability. At one stage it is possible to move only across one edge out of the set E. The searcher is assumed to know the initial position of the mobile targets. Each next position of the hiders is unknown to either the searcher or the other mobile object. The game proceeds from step k to step k-1. The mobile objects do not return to the vertexes they have already been to.

Let $G = \langle V, E \rangle$ be a fixed-root directed tree without loops, where V is the vertex set, E is the edge set. All vertices of the graph are numbered; the root of the tree is denoted by zero. $E = \{(i, j)\}, i, j \in V$, where $(i, j) \in E$ is a directed edge of the graph G. If $(i, j) \in E$, then vertex j is called descendant of vertex i. Denote by L the set of leaf vertices G. Define for any vertex j the offspring set $ch(j)$ of vertex j. We imply two restrictions for the set E: 1. $\forall j \in V, j \neq 0 \exists! i \in V : (i, j) \in E$; 2. $(i, i) \notin E$.

The vector of vertices $(i_1, i_2, \ldots, i_k)$, where $\forall j = 1, 2, \ldots, k - 1 : (i_j, i_{j+1}) \in E$, will be termed the path in the graph G. The length of any given edge is 1, $|(i_1, i_2, \ldots, i_k)| = k - 1$ is the length of the path $(i_1, i_2, \ldots, i_k)$, $\forall (i, j) \in E |(i, j)| = 1$. Let there be $(0, i_1, i_2, \ldots, i_k), (0, j_1, \ldots, j)$ so we can infer that $\forall i, j \in L : |(0, i_1, \ldots, i)| = |(0, j_1, \ldots, j)| = n$ is the maximum length of a path in the graph G. In a multi-stage game the number n is called the number of stages in the multi-stage game. $v(k) = \{i | i \in V, \exists i_2, \ldots, i_k \in V : |(0, i_2, \ldots, i_k, i)| = k\}$ is the number of vertices situated at distance k from the root.

Image the following situation. It suffices for the searcher to locate at least one mobile object, after which the search stops. The mobile objects seek to minimize their own probability of being captured.

At stage 1 (last stage), let the first mobile object occupy vertex g and the second one vertex l; $g, l \in v(n - 1), g \neq l$. Define the game

$$\Gamma(1, g, l, \Phi_g, \Phi_l) = \langle I, II, III; P(g), Q(l), \Psi(g) \cup \Psi(l); H_1(\cdot), H_2(\cdot), H_3(\cdot) \rangle,$$

where I, II, III are the players. $P(g), Q(l), \Psi(g) \cup \Psi(l)$ are the sets of the players' strategies. Let us determine the set of strategies for the first mobile object.

$$\forall g \in V \setminus L : P(g) = \left\{ (p_i)_{i \in ch(g)} | p_i \geq 0, \sum_{i \in ch(g)} p_i = 1 \right\},$$

where p_i is the probability that the first mobile object will move from the ancestor of vertex i to vertex i. $(p_i)_{i \in ch(g)}$ is the vector of dimensionality $ch(g)$. If $\exists i \in ch(g) : p_i = 1$, then $(p_i)_{i \in ch(g)}$ is the first player's pure strategy, otherwise it is mixed. $P(g)$ is the set of strategies of the first player when at vertex g. For brevity, the notation $p(g) = (p_i)_{i \in ch(g)}$ is introduced. Let us now describe the set of strategies for the second player.

$$\forall l \in V \setminus L : \quad Q(l) = \left\{ (q_j)_{j \in ch(l)} | q_l \geq 0, \sum_{j \in ch(l)} q_j = 1 \right\},$$

where q_j is the probability that the second mobile object will move from the ancestor of vertex j to vertex j. $(q_j)_{j \in ch(l)}$ is a vector of dimensionality $h(g)$. $Q(l)$ is the set of strategies of the second player when at vertex l. For brevity, the notation $q(l) = (q_j)_{j \in ch(l)}$ is introduced.

Denote by Φ the searcher's budget. Φ_g is the share of Phi distributed among the vertices with the parent g. The budget allocated to a clique is distributed among vertices in the clique, i.e. $\Phi = \sum_{g \in V \setminus B} \Phi_g$, $\sum_{i \in ch(g)} c_i \varphi_i = \Phi_g \geq \phi_g > 0$. A game model with a distribution of resource is presented in [8]. Below is the set of strategies for the searcher.

$$\forall g \in V \setminus L : \quad \Psi(g) = \left\{ (\varphi_i)_{i \in ch(g)} | \varphi_i \geq 0; \sum_{i \in ch(g)} c_i \varphi_i = \Phi_g \right\},$$

where $\varphi_i, i \in ch(g)$ is the searching resource allocated for the retrieval of a hidden object to vertex i; $(\varphi_i)_{i \in ch(g)}$ is the vector from $\varphi_i, i \in ch(g)$ and the searcher's strategy; $c_i > 0$ is the cost of a unit resource allocated to vertex i; $\sum_{i \in ch(g)} c_i \varphi_i$ is the costs the searcher spent on detecting the mobile objects among descendants of vertex j. $\Psi(g)$ is the set of strategies for the searcher when one or both mobile objects are at vertex g; $\Psi(g) \cup \Psi(l)$ is the set of strategies for the searcher when the first mobile object is at vertex g, while the second one is at vertex l. For brevity, the notation $\varphi(g) = (\varphi_i)_{i \in ch(g)}$ is introduced.

Suppose that the detection probability of, respectively, the first or the second mobile object at vertex i is $1 - e^{-\alpha_i \varphi_i}$, provided that the mobile object is at vertex i. The searching capacity of the resource φ_i is determined by the factor $\alpha_i > 0$. The probability of not detecting, respectively, the first or the second mobile object at vertex i then equals $e^{-\alpha_i \varphi_i}$.

2 Single-Stage Game

Let us determine the players' payoff functions at stage 1.

$$H_1(1, p(g), \varphi(g)) = \sum_{i \in ch(g)} p_i (1 - e^{-\alpha_i \varphi_i}),$$

$$H_2(1, q(l), \varphi(l)) = \sum_{j \in ch(l)} q_j(1 - e^{-\alpha_j \varphi_j}),$$

$$H_3(1, p(g), q(l), \varphi(g), \varphi(l)) = \sum_{i \in ch(g)} \sum_{j \in ch(l)} p_i q_j(1 - e^{-\alpha_i \varphi_i - \alpha_j \varphi_j}),$$

$$p(g) \in P(g), q(l) \in Q(l), \varphi(g) \in \Psi(g), \varphi(l) \in \Psi(l), \tag{1}$$

$$\forall i \in ch(g), \forall j \in ch(l) : p_i \geq 0, q_j \geq 0, \varphi_i \geq 0, \varphi_j \geq 0, \tag{2}$$

$$\sum_{i \in ch(g)} p_i = 1, \sum_{j \in ch(l)} q_j = 1, \sum_{i \in ch(g)} c_i \varphi_i = \Phi_g, \sum_{j \in ch(l)} c_j \varphi_j = \Phi_l, \tag{3}$$

where $H_1(\cdot)$, $H_2(\cdot)$, $H_3(\cdot)$ is the probability of detecting the first, the second, and at least one mobile object, respectively. The first mobile object moves from vertex g to a new vertex i, $(g, i) \in E$ with probability $p_i \geq 0$, $\sum_{i \in ch(g)} p_i = 1$. Similarly, the second mobile object moves from vertex l to j, $(l, j) \in E$ with probability $q_j \geq 0$, $\sum_{j \in ch(l)} q_l = 1$.

The searcher, in turn, allocates a budget of Φ_g, Φ_l to the search for at least one mobile object, with Φ_g allocated for searching among descendants of vertex g, and Φ_l for searching among descendants of vertex l. The budget Φ_g is used to purchase resources $\varphi_i \geq 0$, $i \in ch(g)$ at $c_i > 0$ per unit resource, and all of Φ_g is spent. Likewise, the budget Φ_l is used to purchase resources $\varphi_j \geq 0$, $j \in ch(l)$ at $c_j > 0$ per unit resource, and Φ_l is fully spent. Since the first mobile object moves to vertex i with probability p_i, and the searcher channels φ_i resources to i, hence $p_i(1 - e^{-\alpha_i \varphi_i})$ is the probability of detecting the first mobile object at vertex i; $q_j(1 - e^{-\alpha_j \varphi_j})$ is the probability of detecting the second mobile object at vertex j; $p_i q_j(1 - e^{-\alpha_i \varphi_i - \alpha_j \varphi_j})$ is the probability of detecting at least one mobile object in the situation where the first mobile object has moved to vertex i, and the second one to vertex j.

All components of the game $\Gamma(1, g, l, \Phi_n)$ have now been defined for the situation where the mobile objects are situated at different vertices. If at stage 1 the mobile objects both occupy the same vertex g, the following notations are introduced:

$$\Gamma(1, g, \Phi_g) = \langle I, II, III; P(g), Q(g), \Psi(g); H_1(\cdot), H_2(\cdot), H_3(\cdot) \rangle,$$

$$H_1(1, p(g), \varphi(g)) = \sum_{i \in ch(g)} p_i(1 - e^{-\alpha_i \varphi_i}),$$

$$H_2(1, q(g), \varphi(g)) = \sum_{j \in ch(g)} q_j(1 - e^{-\alpha_j \varphi_j}),$$

Fig. 1 Graph G

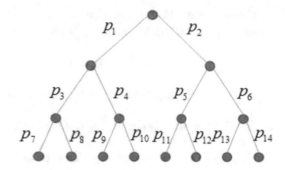

$$H_3(1, p(g), q(g), \varphi(g)) = \sum_{i \in ch(g)} \sum_{j \in ch(g)} p_i q_j (1 - e^{-\alpha_i \varphi_i - \alpha_j \varphi_j}),$$

$$p(g) \in P(g), q(l) \in Q(l), \varphi(g) \in \Psi(g), \tag{4}$$

$$\forall i, j \in ch(g) : p_i \geq 0, q_j \geq 0, \varphi_i \geq 0, \varphi_j \geq 0, \tag{5}$$

$$\sum_{i \in ch(g)} p_i = 1, \quad \sum_{j \in ch(g)} q_j = 1, \quad \sum_{i \in ch(g)} c_i \varphi_i = \Phi_g. \tag{6}$$

Example 1 Consider the graph in Fig. 1. Probability $p_i \geq 0, i \in V \setminus \{0\}$ is written at each edge. It denotes the probability of the first mobile object moving to vertex i from the ancestor of vertex i. Then $p_{2j-1} + p_{2j} = 1, j = 1, 2, \ldots, 7$. Similarly, $q_i \geq 0, i \in V \setminus \{0\}$ is the probability of the second mobile object moving to vertex i from the ancestor of vertex i. Then $q_{2j-1} + q_{2j} = 1, j = 1, 2, \ldots, 7$. The following equalities are true:

$$ch(0) = \{1, 2\}, ch(1) = \{3, 4\}, ch(2) = \{5, 6\}, ch(3) = \{7, 8\},$$

$$ch(4) = \{9, 10\}, ch(5) = \{11, 12\}, ch(6) = \{13, 14\},$$

$$B = \{7, 8, \ldots, 14\}, v(1) = \{1, 2\}, v(2) = \{3, 4, 5, 6\}, v(3) = \{7, 8, \ldots, 14\}.$$

$$\sum_{i=0}^{6} \Phi_i = \Phi, c_1 \varphi_1 + c_2 \varphi_2 = \Phi_0, c_3 \varphi_3 + c_4 \varphi_4 = \Phi_1, c_5 \varphi_5 + c_6 \varphi_6 = \Phi_2,$$

$$c_7 \varphi_7 + c_8 \varphi_8 = \Phi_3, c_9 \varphi_9 + c_{10} \varphi_{10} = \Phi_4, c_{11} \varphi_{11} + c_{12} \varphi_{12} = \Phi_5,$$

$$c_{13} \varphi_{13} + c_{14} \varphi_{14} = \Phi_6.$$

If there exists a Nash equilibrium in the game $\Gamma_2(1, g, l, \Phi_g, \Phi_l)$, then there are strategies $p^*(g), q^*(l), \varphi^*(g), \varphi^*(l)$, such that the inequalities

$$H_1(1, p(g), \varphi^*(g)) \geq H_1(1, p^*(g), \varphi^*(g)) \forall p(g) \in P(g),$$

$$H_2(1, q(l), \varphi^*(l)) \geq H_2(1, q^*(l), \varphi^*(l)) \forall q(l) \in Q(l),$$

$$H_3(1, p^*(g), q^*(l), \varphi(g), \varphi^*(l)) \leq H_3(1, p^*(g), q^*(l), \varphi^*(g), \varphi^*(l)) \forall \varphi(g) \in \Psi(g),$$

$$H_3(1, p^*(g), q^*(l), \varphi^*(g), \varphi(l)) \leq H_3(1, p^*(g), q^*(l), \varphi^*(g), \varphi^*(l)) \forall \varphi(l) \in \Psi(l),$$

hold, so denote $H_1^*(1, g, \Phi_g) = H_1(1, p^*(g), \varphi^*(g))$, $H_2^*(1; l, \Phi_l) = H_2(1, q^*(l), \varphi^*(l))$,
$H_3^*(1, g, l, \Phi_g, \Phi_l) = H_3(1, p^*(g), q^*(l), \varphi^*(g), \varphi^*(l))$.

Similarly, in the game $\Gamma_1(1, g, \Phi_g)$ we are interested in the Nash equilibrium, i.e. such players' strategies $p^*(g), q^*(q), \varphi^*(g)$ for which the inequalities

$$H_1(1, p(g), \varphi^*(g)) \geq H_1(1, p^*(g), \varphi^*(g)) \forall p(g) \in P(g),$$

$$H_2(1, q(g), \varphi^*(g)) \geq H_2(1, q^*(g), \varphi^*(g)) \forall q(g) \in Q(g),$$

$$H_3(1, p^*(g), q^*(g), \varphi(g)) \leq H_3(1, p^*(g), q^*(l), \varphi^*(g)) \forall \varphi(g) \in \Psi(g),$$

hold. Denote $H_3^*(1, g, \Phi_g) = H_3(1, p^*(g), q^*(g), \varphi^*(g))$.

3   Multi-stage Game

Let us now define the game $\Gamma_2(k, g, l, \Phi_g, \Phi_l)$, $\Gamma_1(k, g, \Phi_g)$ at stages $2, 3, \ldots, n$. Let at stage $k = 2, 3, \ldots, n$ the first and the second mobile objects occupy vertices $g, l \in v(n - k - 1)$, respectively.

$$\Gamma_2(k, g, l, \Phi_g, \Phi_l) = \langle I, II, III; P(g), Q(l), \Psi(g) \cup \Psi(l); H_1(\cdot), H_2(\cdot), H_3(\cdot) \rangle .$$

The players' payoff functions have the form

$$H_1(k, p(g), \varphi(g)) = \sum_{i \in ch(g)} p_i(1 - e^{-\alpha_i \varphi_i}) + \sum_{i \in ch(g)} p_i e^{-\alpha_i \varphi_i} H_1^*(k - 1, i, \Phi_i),$$

$$H_2(k, q(l), \varphi(l)) = \sum_{j \in ch(l)} q_j(1 - e^{-\alpha_j \varphi_j}) + \sum_{j \in ch(l)} q_j e^{-\alpha_j \varphi_j} H_2^*(k - 1, j, \Phi_j),$$

$$H_3(k, p(g), q(l), \varphi(g), \varphi(l)) = \sum_{i \in ch(g)} \sum_{j \in ch(l)} p_i q_j (1 - e^{-\alpha_i \varphi_i - \alpha_j \varphi_j}) +$$

$$+ \sum_{i \in ch(g)} \sum_{j \in ch(j)} p_i q_j e^{-\alpha_i \varphi_i - \alpha_j \varphi_j} H_3^*(k - 1, i, j, \Phi_i, \Phi_j).$$

The variables are bound by restrictions (1)–(3). Let at stage $k = 2, 3, \ldots, n - 1$ both mobile objects occupy the same vertex $g \in v(n - k + 1)$. Consider the game

$$\Gamma_1(k, g, \Phi_k) = \langle I, II, III; P(g), Q(q), \Psi(g); H_1(\cdot), H_2(\cdot), H_3(\cdot) \rangle$$

at this stage. The players' payoff functions have the form

$$H_1(k, p(g), \varphi(g)) = \sum_{i \in ch(g)} p_i (1 - e^{-\alpha_i \varphi_i}) + \sum_{i \in ch(g)} p_i e^{-\alpha_i \varphi_i} H_1^*(k - 1, i, \Phi_i),$$

$$H_2(k, q(g), \varphi(g)) = \sum_{j \in ch(g)} p_j (1 - e^{-\alpha_j \varphi_j}) + \sum_{j \in ch(g)} p_j e^{-\alpha_j \varphi_j} H_2^*(k - 1, j, \Phi_j),$$

$$H_3(k, p(g), q(g), \varphi(g)) = \sum_{i \in ch(g)} \sum_{j \in ch(g)} p_i q_j (1 - e^{-\alpha_i \varphi_i - \alpha_j \varphi_j}) +$$

$$+ \sum_{i \in ch(g)} p_i q_i e^{-2\alpha_i \varphi_i} H_3^*(k - 1, i, \Phi_i) +$$

$$+ \sum_{i \in ch(g)} \sum_{j \in ch(g), i \neq j} p_i q_j e^{-\alpha_i \varphi_i - \alpha_j \varphi_j} H_3^*(k - 1, i, j, \Phi_i, \Phi_j).$$

The variables are bound by restrictions (4)–(6).

If there exists a Nash equilibrium in the game $\Gamma_1(k, g, \Phi_k)$, then denote $H_1^*(k, g, \Phi_g) = H_1(k, p^*(g), \varphi^*(g))$, $H_2^*(k, g, \Phi_g) = H_2(k, q^*(g), \varphi^*(g))$, $H_3^*(k, g, \Phi_g) = H_3(k, p^*(g), q^*(g), \varphi^*(g))$.

If there exists a Nash equilibrium in the game $\Gamma_2(k, g, l, \Phi_g, \Phi_l)$ and the respective optimal strategies for the players coincide, then denote $H_3^*(k, g, l, \Phi_g, \Phi_l) = H_3(k, p^*(g), q^*(l), \varphi^*(g), \varphi^*(l))$, where $(p^*(g), q^*(g), \varphi^*(g))$ is the equilibrium in the game $\Gamma_2(k, g, l, \Phi_g, \Phi_l)$.

4 Existence of Equilibrium

Before looking for equilibrium one has to be sure it exists. The most popular tool for proving the existence of equilibrium is the Nash theorem, for which convexity of payoff functions is essential.

Claim 1 The function $f(x_1, \ldots, x_r) = \sum_{i=1}^{r} \sum_{j=1}^{r} b_{ij} e^{-a_i x_i - a_j x_j}$ is convex downward on the set $D = \{(x_1, \ldots, x_r) | x_i \geq 0, \sum_{i=1}^{r} c_i x_i = \Phi\}$, $a_i, c_i > 0, b_{ij} \in [0; 1]$.

Proof It is obvious that the function $y = e^x$ is convex downward on the set $\in (-\infty; +\infty)$, then $e^{\alpha x_1 + (1-\alpha) x_2} \leq \alpha e^{x_1} + (1 - \alpha) e^{x_2}, \forall \alpha \in [0; 1], \forall x_1, x_2 \in (-\infty; +\infty)$.

Let $x = (x_1, \ldots, x_r), y = (y_1, \ldots, y_r)$. One can show that $\forall x, y \in D$ holds $f(\alpha x + (1 - \alpha) y) \leq \alpha f(x) + (1 - \alpha) f(y), \alpha \in [0; 1]$

$$f(\alpha x + (1 - \alpha) y) = f(\alpha x_1 + (1 - \alpha) y_1, \ldots, \alpha x_r + (1 - \alpha) y_r) =$$

$$= \sum_{i=1}^{r} \sum_{i=1}^{r} b_{ij} e^{-a_i (\alpha x_i + (1-\alpha) y_i) - a_j (\alpha x_j + (1-\alpha) y_j)} =$$

$$= \sum_{i=1}^{r} \sum_{i=1}^{r} b_{ij} e^{\alpha(-a_i x_i - a_j x_j) + (1-\alpha)(-a_j y_j - a_j y_j)}$$

$$\leq \sum_{i=1}^{r} \sum_{i=1}^{r} b_{ij} (\alpha e^{-a_i x_i - a_j x_j} + (1 - \alpha) e^{-a_i y_i - a_j y_j}) =$$

$$= \alpha f(x) + (1 - \alpha) f(y).$$

The above equalities and estimates hold for $\forall x, y \in D$. Hence, $f(x)$ in D is convex downward by definition. □

Lemma 1 *There exists a Nash equilibrium in the games* $\Gamma_1(k, g, \Phi_g)$ *and* $\Gamma_2(k, g, l, \Phi_g, \Phi_l)$.

Proof The functions $H_1(k, p(g), \varphi(g)), H_2(k, q(g), \varphi(g))$ are linear on $p_i, q_i, i \in ch(g)$, respectively. $-H_3(k, p(g), q(g), \varphi(g))$ is convex upward. Since the players' payoff functions are convex and continuous, and the set of strategies is a compact convex set then, according to the Nash theorem, there exists an equilibrium solution. □

5 Solution for a Single-Stage Game

Now that we know equilibrium exists (Lemma 1), let us find the value of the game for a single-stage game by non-linear programming methods.

Theorem 1 *The following equalities are true in the game* $\Gamma_2(1, g, l, \Phi_g, \Phi_l)$

$$H_1^*(1, g, \Phi_g) = 1 - exp\left(\frac{-\Phi_g}{\sum_{i \in ch(g)} \frac{c_i}{\alpha_i}}\right), \quad H_2^*(1, l, \Phi_l) = 1 - exp\left(\frac{-\Phi_l}{\sum_{j \in ch(l)} \frac{c_j}{\alpha_j}}\right),$$

$$H_3^*(1, g, l, \Phi_j, \Phi_l) = 1 - exp\left(-\frac{\Phi_g}{\sum_{i \in ch(g)} \frac{c_i}{\alpha_i}} - \frac{\Phi_l}{\sum_{j \in ch(l)} \frac{c_j}{\alpha_j}}\right).$$

Proof Write down the Kuhn–Tucker conditions.

$$
\begin{cases}
1 - e^{-\alpha_i \varphi_i} - u_i + \lambda_1 = 0, i \in ch(g); \\
1 - e^{-\alpha_j \varphi_j} - s_j + \lambda_2 = 0, j \in ch(l); \\
\alpha_i \sum_{j \in ch(l)} p_i q_j e^{-\alpha_i \varphi_i - \alpha_j \varphi_j} - w_i + c_i \lambda_3 = 0, i \in ch(g); \\
\alpha_j \sum_{i \in ch(g)} p_i q_j e^{-\alpha_i \varphi_i - \alpha_j \varphi_j} - w_j + c_j \lambda_4 = 0, j \in ch(l); \\
\sum_{i \in ch(g)} p_i = 1; \\
\sum_{j \in ch(l)} q_j = 1; \\
\sum_{i \in ch(g)} c_i \varphi_i = \Phi_g; \\
\sum_{j \in ch(l)} c_j \varphi_j = \Phi_l; \\
p_i u_i = 0; \\
q_j v_j = 0; \\
\varphi_i w_i = 0; \\
\varphi_j w_j = 0; \\
p_i, q_j, \varphi_i, \varphi_j, u_i, v_j, w_i, w_j \geq 0, i \in ch(g), j \in ch(l).
\end{cases}
$$

Suppose that $p_i, q_j, \varphi_i, \varphi_j > 0$. Then $u_i = s_j = w_i = w_j = 0$ $\forall i \in ch(g)$, $j \in ch(l)$. Having solved the system of equations were get that $p_i^* = \frac{c_i/\alpha_i}{\sum_{m \in ch(g)} c_m/\alpha_m}, q_j^* = \frac{c_j/\alpha_j}{\sum_{m \in ch(l)} c_m/\alpha_m}, \varphi_i^* = \frac{\Phi_g/\alpha_i}{\sum_{m \in ch(g)} c_m/\alpha_m}, \varphi_j^* = \frac{\Phi_l/\alpha_j}{\sum_{m \in ch(l)} c_m/\alpha_m}$.
Since the resultant transition probabilities of the mobile players and the numerical value of the resource are non-negative, there is no need to consider other cases of the sign of the variables u_i, s_j, w_i. Substituting the resultant values of the variables into the players' payoff functions we get the required proof. □

Theorem 2 *The following equalities are true in the game* $\Gamma_1(1, 0, \Phi)$

$$
H_1^*(1, 0, \Phi) = H_2^*(1, 0, \Phi) = 1 - exp\left(\frac{-\Phi}{\sum_{i \in ch(0)} \frac{c_i}{\alpha_i}}\right), H_3^*(1, 0, \Phi) = 1 - exp\left(\frac{-2\Phi}{\sum_{i \in ch(0)} \frac{c_i}{\alpha_i}}\right).
$$

Proof Write down the Kuhn–Tucker conditions.

$$
\begin{cases}
1 - e^{-\alpha_i \varphi_i} - u_i + \lambda_1 = 0, i \in ch(0); \\
1 - e^{-\alpha_i \varphi_i} - v_i + \lambda_2 = 0, i \in ch(0); \\
2\alpha_i p_i q_i e^{-2\alpha_i \varphi_i} + \alpha_i \sum_{j \in ch(0), i \neq j} (p_i q_j + p_j q_i) e^{-\alpha_i \varphi_i - \alpha_j \varphi_j} - \\
\quad - w_i + c_i \lambda_3 = 0, i \in ch(0); \\
\sum_{i \in ch(0)} p_i = 1; \\
\sum_{i \in ch(0)} q_i = 1; \\
\sum_{i \in ch(0)} c_i \varphi_i = \Phi; \\
p_i u_i = 0; \\
q_i v_i = 0; \\
\varphi_i w_i = 0; \\
p_i, q_i, \varphi_i, u_i, v_i, s_i \geq 0, i \in ch(0).
\end{cases}
$$

Suppose that $p_i, q_i, \varphi_i > 0$. Then $u_i = s_i = w_i = 0 \forall i \in ch(0)$. Then from the first $2|ch(0)|$)l equations of the system we get $\lambda_1 = \lambda_2, \varphi_i = -\frac{\ln(\lambda+1)}{\alpha_i}$. Substituting the resultant values of φ_i into $\sum_{i \in ch(0)} c_i \varphi_i = \Phi$ we find that $\varphi_i^* = \frac{\Phi/\alpha_i}{\sum_{j \in ch(0)} c_j/\alpha_j}$. Substitute φ_i^*hi into the remaining equations of this system.

$$\frac{\alpha_i}{c_i} exp\left(-\frac{2\Phi}{\sum_{j \in ch(0)} c_i/\alpha_i}\right)\left(2p_i q_i + \sum_{j \in ch(0)} (p_i q_j + p_j q_i)\right) + \lambda_3 = 0;$$

$$\frac{\alpha_i}{c_i}(p_i + q_i) = -\lambda_3 \cdot exp\left(-\frac{2\Phi}{\sum_{j \in ch(0)} c_i/\alpha_i}\right).$$

The right-hand part of the equation is independent of i, hence $\forall i, j \in ch(0)$: $\frac{\alpha_i}{c_i}(p_i + q_i) = \frac{\alpha_j}{c_j}(p_j + q_j)$. Since the sum of probabilities is 1, we find that $p_i^* = \gamma_i, q_i^* = 2 \cdot \frac{c_i/\alpha_i}{\sum_{j \in ch(0)} c_j/\alpha_j} - \gamma_i$, where $0 \le \gamma_i \le 2 \cdot \frac{c_i/\alpha_i}{\sum_{j \in ch(0)} c_j/\alpha_j}$, $\sum_{j \in ch(0)} \gamma_j = 1$. Substituting the resultant values of $p_i^*, q_i^*, \varphi_i^*$ into the payoff functions we get the required proof. Note that the optimal strategies of the players are interdependent. Since the mobile objects do not know exactly whether they share the same vertex, the players will choose such $\gamma_i, i \in ch(0)$ for which $p_i^* = q_i^* = \frac{c_i/\alpha_i}{\sum_{j \in ch(0)} c_j/\alpha_j}$ holds. $\qquad\square$

Example 2 Let us solve a single-stage game for the graph G shown in Fig. 2. The values of the parameters c_i, α_i are set in Table 1.

The mobile players were initially deployed at vertex 0. Their next positions could be either in the same or in different vertices (both in the first, both in the second, one in the first, and the other one in the second). It follows from Theorem 3 that the searcher need not know how many mobile players there are in the 1st and 2nd vertices to distribute the searching resource, he just allocates a budget of $\varphi_i = \frac{\Phi_g/\alpha_i}{\sum_{m \in ch(g)} c_m/\alpha_m}, i \in ch(g), g \in \{1, 2\}$ to vertex i. For the numerical value of φ see Table 2. For mobile objects occupying different vertices or the same

Fig. 2 Graph G

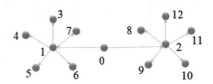

Table 1 Values of parameters c_i, α_i

i	1	2	3	4	5	6	7	8	9	10	11	12
α_i	0.9	0.3	0.4	0.5	0.8	0.2	0.6	0.5	0.7	0.2	0.8	0.5
c_i	7	4	2	6	8	2	5	3	4	1	8	5

Table 2 Values of p_i, φ_i

i	3	4	5	6	7	8	9	10	11	12
p_i	0.110	0.265	0.220	0.220	0.185	0.163	0.155	0.136	0.273	0.273
φ_i	5514	4411	2757	11,029	3676	10,894	7782	27,237	6809	10,893

vertex their probabilities of transition to vertex i are calculated by the formula $p_i = \frac{c_i/\alpha_i}{\sum_{m \in ch(g)} c_m/\alpha_m}$, as shown in Table 2.

6 Solution of the Game for a Linear Graph

Let L_{n+1} be a linear graph with $n+1$ vertices, where n is the number of stages. We assume that at stage k the mobile objects are at vertex $k+1$, $k = n, n-1, \ldots, 1$. Initially, both mobile objects are at vertex $n+1$. Since the graph is linear, the players move to the next vertex with probability 1. For convenience we assume that a budget of Φ_i, $i = 1, 2, \ldots, n$ is allocated to the ith vertex.

Theorem 3 *In the game* $\Gamma_1(n, n+1, \Phi_n)$ $H_3^*(n, n+1, \Phi_n) = 1 - e^{-2\sum_{i=1}^{n} \frac{\alpha_i}{c_i} \Phi_i}$, $H_1^*(n, n+1, \Phi_n) = H_2^*(n, n+1, \Phi_n) = 1 - e^{-\sum_{i=1}^{n} \frac{\alpha_i}{c_i} \Phi_i}$ *holds.*

Proof Note that for a linear graph vectors $p(k)$, $q(k)$ consist of one component p_{k-1}, q_{k-1}, respectively.

$$H_3(1, p_1, q_1, \varphi_1) = p_1 q_1 (1 - e^{-\alpha_1 \varphi_1}).$$

Since $p_1 = q_1 = 1$, $c_1 \varphi_1 = \Phi_1$, then

$$H_3^*(1, 2, \Phi_1) = 1 - e^{-\frac{\alpha_1}{c_1} \Phi_1},$$

$$H_3(2, p_2, q_2, \varphi_2) = p_2 q_2 (1 - e^{-\alpha_2 \varphi_2}) + p_2 q_2 e^{-\alpha_2 \varphi_2} H_3^*(1, 2, \Phi_1),$$

$$H_3^*(2, 3, \Phi_2) = 1 - e^{-2\frac{\alpha_2}{c_2} \Phi_2} + e^{-2\frac{\alpha_2}{c_2} \Phi_2}(1 - e^{-2\frac{\alpha_1}{c_1} \Phi_1}) = 1 - e^{-2(\frac{\alpha_2}{c_2} \Phi_2 + \frac{\alpha_1}{c_1} \Phi_1)}.$$

At the kth stage we have

$$H_3(k, p_k, q_k, \varphi_k) = p_k q_k (1 - e^{-\alpha_k \varphi_k}) + p_k q_k e^{-\alpha_k \varphi_k} H_3^*(k-1, k, \Phi_{k-1})$$

$$H_3^*(k, k+1, \Phi_k) = 1 - e^{-2\frac{\alpha_k}{c_k} \Phi_k} + e^{-2\frac{\alpha_k}{c_k} \Phi_k}(1 - e^{-2\sum_{i=1}^{k-1} \frac{\alpha_i}{c_i} \Phi_i}) = 1 - e^{-2\sum_{i=1}^{k} \frac{\alpha_i}{c_i} \Phi_i}.$$

When $k = n$ we get $H_3^*(n, n + 1, \Phi_n) = 1 - e^{-2\sum_{i=1}^{n} \frac{\alpha_i}{c_i} \Phi_i}$. Following the same reasoning we can find that $H_1^*(n, n+1, \Phi_n) = H_2^*(n, n+1, \Phi_n) = 1 - e^{-\sum_{i=1}^{n} \frac{\alpha_i}{c_i} \Phi_i}$.

$\square$

Knowing the probability of detecting at least one mobile object within n stages, the searcher can distribute the budget Φ optimally, e.g. by maximizing $H_3^*(n, n + 1, \Phi_n)$ using the variables $\Phi_i \geq 0, i = 1, 2, \ldots, n, \sum_{i=1}^{n} \Phi_i = \Phi$.

Claim 2 The highest probability of detecting at least one mobile object on a linear graph is $H_3^*(n, n + 1, \Phi) = 1 - e^{-2\Phi \max_{i=1,\ldots,n} \frac{\alpha_i}{c_i}}$

Proof $1 - e^{-2\sum_{i=1}^{n} \frac{\alpha_i}{c_i} \Phi_i} \to max \Leftrightarrow -e^{-2\sum_{i=1}^{n} \frac{\alpha_i}{c_i} \Phi_i} \to max \Leftrightarrow e^{-2\sum_{i=1}^{n} \frac{\alpha_i}{c_i} \Phi_i} \to min \Leftrightarrow -2\sum_{i=1}^{n} \frac{\alpha_i}{c_i} \Phi_i \to min \Leftrightarrow \sum_{i=1}^{n} \frac{\alpha_i}{c_i} \Phi_i \to max$. Let $\frac{\alpha_j}{c_j} = \max_{i=1,\ldots,n} \frac{\alpha_i}{c_i}$.

We deem the value of j to be fixed. Since we find the maximum of the linear function with the restrictions $\Phi_i \geq 0, i = 1, 2, \ldots, n, \sum_{i=1}^{n} \Phi_i = \Phi$, so $\Phi_i = \begin{cases} 0, i \neq j; \\ \Phi, i = j. \end{cases}$, and then $H_3(n, n + 1, \Phi) = 1 - e^{-2\Phi \max_{i=1,\ldots,n} \frac{\alpha_i}{c_i}}$, as was to be proved.

In other words, for the detection probability to be the highest possible, the entire budget is allocated to a single vertex for which the fraction $\frac{\alpha_i}{c_i}, i \in V \setminus \{n\}$ is the highest.

$\square$

7 Solution for a Two-Stage Game

It follows from Theorems 3 and 5 that in a single-stage game the optimal probability of detecting the first and the second mobile objects, respectively, is $1 - 1 - e^{-\frac{\Phi_g}{\sum_{i \in ch(g)} c_i / \alpha_i}}$, where g is the vertex in which the mobile object is situated. Denote $\beta_g = e^{-\frac{\Phi_g}{\sum_{i \in ch(g)} c_i / \alpha_i}}$. Then $H_1^*(1, g, \Phi_g) = H_2^*(1, g, \Phi_g) = 1 - \beta_g, H_3^*(1, g, \Phi_g) = 1 - \beta_g^2, H_3^*(1, g, l, \Phi_g, \Phi_l) = 1 - \beta_g \beta_l$.

Let the mobile objects at the second stage be at vertex $g, g \in v(n - 2)$, then the players' payoff functions in the game $\Gamma_1(2, g, \Phi_g) = \langle I, II, III; P(g), Q(q), \Psi(g); H_1(\cdot), H_2(\cdot), H_3(\cdot) \rangle$ will have the form

$$H_1(2, p(g), \varphi(g)) = \sum_{i \in ch(g)} p_i(1 - e^{-\alpha_i \varphi_i}) + \sum_{i \in ch(g)} p_i e^{-\alpha_i \varphi_i}$$

$$H_1^*(1, i, \Phi_i) = 1 - \sum_{i \in ch(g)} p_i e^{-\alpha_i \varphi_i} + \sum_{i \in ch(g)} p_i e^{-\alpha_i \varphi_i}(1 - \beta_i) = 1 - \sum_{i \in ch(g)} \beta_i p_i e^{-\alpha_i \varphi_i},$$

$$H_2(2, p(g), \varphi(g)) = 1 - \sum_{i \in ch(g)} \beta_i q_i e^{-\alpha_i \varphi_i}, \quad H_3(2, p(g), q(g), \varphi(g)) =$$

$$\sum_{i \in ch(g)} \sum_{j \in ch(g)} p_i q_j (1 - e^{-\alpha_i \varphi_i - \alpha_j \varphi_j}) + \sum_{i \in ch(g)} p_i q_i e^{-2\alpha_i \varphi_i} H_3^*(k-1, i, \Phi_i)$$

$$+ \sum_{i \in ch(g)} \sum_{j \in ch(g), i \neq j} p_i q_j e^{-\alpha_i \varphi_i - \alpha_j \varphi_j} H_3^*(k-1, i, j, \Phi_i, \Phi_j) =$$

$$= 1 - \sum_{i \in ch(g)} \sum_{j \in ch(g)} p_i q_j e^{-\alpha_i \varphi_i - \alpha_j \varphi_j} + \sum_{i \in ch(g)} p_i q_i e^{-2\alpha_i \varphi_i}(1 - \beta_i^2) +$$

$$+ \sum_{i \in ch(g)} \sum_{j \in ch(g), i \neq j} p_i q_j e^{-\alpha_i \varphi_i - \alpha_j \varphi_j}(1 - \beta_i \beta_j) =$$

$$= 1 - \sum_{i \in ch(g)} \beta_i^2 p_i q_i e^{-2\alpha_i \varphi_i} - \sum_{i \in ch(g)} \sum_{j \in ch(g), i \neq j} \beta_i \beta_j p_i q_j e^{-\alpha_i \varphi_i - \alpha_j \varphi_j}.$$

Optimal strategies can be found using non-linear programming methods.

Theorem 4 *Optimal strategies in the game* $\Gamma_1(2, g, \Phi_g)$ *have the form*

$$\varphi_i^* = \left[\frac{\Phi_g / \alpha_i}{\sum_{j \subset M(g)} \frac{c_j}{\alpha_j}} - \frac{1}{\alpha_i} \frac{\sum_{j \in M(g)} \frac{c_j}{\alpha_j} ln\beta_j}{\sum_{j \in M(g)} \frac{c_j}{\alpha_j}} + \frac{ln\beta_i}{\alpha_i} \right]^+,$$

$$p_i^* = \begin{cases} \frac{c_i / \alpha_i}{\sum_{j \in M(g)} c_j / \alpha_j}, & i \in M(g); \\ 0, & i \notin M(g). \end{cases}$$

where $M(g)$—*is the number of vertices for which* $\varphi_i > 0, i \in ch(g)$.

Proof Note that the functions $H_1(2, p(g), \varphi(g))$,
$H_2(2, p(g), \varphi(g))$ are symmetric with respect to the variables p_i, q_i, wherefore $p_i = q_i \forall i \in ch(g)$. Write down the Lagrangian functions.

$$L_1(p(g), \varphi^*(g), U, \lambda_1) = 1 - \sum_{i \in ch(g)} \beta_i p_i e^{-\alpha_i \varphi_i^*} - \sum_{i \in ch(g)} p_i u_i + \lambda_1 \left(\sum_{i \in ch(h)} p_i - 1 \right),$$

$$L_2(p^*(g), \varphi(g), W, \lambda_2) = -1 + \sum_{i \in ch(g)} \beta_i^2 (p_i^*)^2 e^{-2\alpha_i \varphi_i} +$$

$$+ \sum_{i \in ch(g)} \sum_{j \in ch(g), i \neq j} \beta_i \beta_j p_i^* p_j^* e^{-\alpha_i \varphi_i - \alpha_j \varphi_j} - \sum_{i \in ch(g)} w_i \varphi_i + \lambda_2 \left(\sum_{i \in ch(g)} c_i \varphi_i - \Phi_g \right).$$

Write down the Kuhn–Tacker conditions required for finding the minimum of the functions $L_1(\cdot)$, $L_2(\cdot)$, and deem $\varphi_i^* = \varphi_i$, $p_i^* = p_i$ for the sake of brevity.

$$
\begin{cases}
-\beta_i e^{-\alpha_i \varphi_i} - u_i + \lambda_1 = 0, i \in ch(g); \\
-2\alpha_i \beta_i^2 p_i^2 e^{-2\alpha_i \varphi_i} - 2\alpha_i \beta_i p_i e^{-\alpha_i \varphi_i} \sum_{j \in ch(g), i \neq j} \beta_j p_j e^{-2\alpha_j \varphi_j} - \\
\qquad -w_i + c_i \lambda_2 = 0, i \in ch(g); \\
p_i u_i = 0; \\
\varphi_i w_i = 0; \\
\sum_{i \in ch(g)} p_i = 1; \\
\sum_{i \in ch(g)} c_i \varphi_i = \Phi_g; \\
p_i, \varphi_i, u_i, w_i \geq 0.
\end{cases}
$$

Transform the system of equations to the form

$$
\begin{cases}
-\beta_i e^{-\alpha_i \varphi_i} - u_i + \lambda_1 = 0, i \in ch(g); \\
-2\alpha_i \beta_i p_i e^{-\alpha_i \varphi_i} \left(\sum_{j \in ch(g)} \beta_j p_j e^{-\alpha_j \varphi_j} \right) - w_i + c_i \lambda_2 = 0, i \in ch(g); \\
p_i u_i = 0; \\
\varphi_i w_i = 0; \\
\sum_{i \in ch(g)} p_i = 1; \\
\sum_{i \in ch(g)} c_i \varphi_i = \Phi_g; \\
p_i, \varphi_i, u_i, w_i \geq 0.
\end{cases}
$$

1. Consider the case where $p_i > 0, \varphi_i > 0 \forall i \in ch(g)$. Then $u_i = w_i = 0 \forall i \in ch(g)$, $\beta_i e^{-\alpha_i \varphi_i} = \lambda_1$, $\varphi_i = -\frac{1}{\alpha_i}(ln\lambda_1 - ln\beta_i)$. Since $\sum_{i \in ch(g)} c_i \varphi_i = \Phi_g$, то $-ln\lambda_1 \sum_{i \in ch(g)} \frac{c_i}{\alpha_i} + \sum_{i \in ch(g)} \frac{c_i}{\alpha_i} ln\beta_i = \Phi_g$,

$$
\varphi_i = \frac{\Phi_g/\alpha_i}{\sum_{j \in ch(g)} \frac{c_j}{\alpha_j}} - \frac{1}{\alpha_i} \frac{\sum_{j \in ch(g)} \frac{c_j}{\alpha_j} ln\beta_j}{\sum_{j \in ch(g)} \frac{c_j}{\alpha_j}} + \frac{ln\beta_i}{\alpha_i}.
$$

Find the values of p_i, $i \in ch(g)$.

$$
-2\alpha_i \beta_i p_i e^{-\alpha_i \varphi_i} \left(\sum_{j \in ch(g)} \beta_j p_j e^{-\alpha_j \varphi_j} \right) - w_i + c_i \lambda_2 = 0;
$$

$$
-2\alpha_i p_i \lambda_1 \left(\sum_{j \in ch(g)} \lambda_1 p_j \right) + c_i \lambda_2 = 0;
$$

$$
2\alpha_i \lambda_1^2 p_i = c_i \lambda_2;
$$

$$\frac{\alpha_i p_i}{c_i} = \frac{\lambda_2}{2\lambda_1^2} \equiv const;$$

$$\frac{\alpha_i p_i}{c_i} = \frac{\alpha_j p_j}{c_j} \forall i, j \in ch(g).$$

Since the $\sum_{i \in ch(g)} p_i = 1$, we get $p_i = \frac{c_i/\alpha_i}{\sum_{j \in ch(g)} c_j/\alpha_j}$

Since the resultant values of φ_i are not always below zero, we also need to consider other cases.

2. Denote by $M(g)$ the set of vertices for which $\varphi_i > 0$. Since Φ_g is positive, $M(g)$ is a non-empty set. Then $\forall i \in M(g) : w_i = 0$. We also assume that $\forall j \in v(n-2) \setminus M(g) : w_j > 0$, then $\varphi_j = 0$. The system of equations then takes the following form:

$$
\begin{cases}
-\beta_i e^{-\alpha_i \varphi_i} - u_i + \lambda_1 = 0, i \in M(g); \\
-\beta_j - u_j + \lambda_1 = 0, j \in v(n-2) \setminus M(g); \\
\frac{\alpha_k \beta_k p_k e^{-\alpha_k \varphi_k}}{c_k} = \frac{\alpha_i \beta_i p_i e^{-\alpha_i \varphi_i}}{c_i}; k, i \in M(g); \\
-2\alpha_j \beta_j p_j \left(\sum_{k \in ch(g)} \beta_k p_k e^{-\alpha_k \varphi_k} \right) - w_j + c_j \lambda_2 = 0; \\
\sum_{i \in ch(g)} p_i = 1; \\
\sum_{i \in ch(g)} c_i \varphi_i = \Phi_g; \\
p_i u_i = 0.
\end{cases}
$$

Let $p_i > 0, i \in M(g), p_j = 0, j \in ch(g) \setminus M(g)$. Then $u_i = 0, i \in M(g)$, We deem $u_j > 0, j \in ch(g) \setminus M(g)$. The system can be rewritten as

$$
\begin{cases}
-\beta_i e^{-\alpha_i \varphi_i} + \lambda_1 = 0, i \in M(g); \\
-\beta_j - u_j + \lambda_1 = 0, j \in ch(g) \setminus M(g); \\
\frac{\alpha_k \beta_k p_k e^{-\alpha_k \varphi_k}}{c_k} = \frac{\alpha_i \beta_i p_i e^{-\alpha_i \varphi_i}}{c_i}, k, i \in M(g); \\
w_j = c_j \lambda_2; \\
\sum_{i \in M(g)} p_i = 1; \\
\sum_{i \in M(g)} c_i \varphi_i = \Phi_g.
\end{cases}
$$

From the equation $\beta_i e^{-\alpha_i \varphi_i} = \lambda_1$ we get that $\varphi_i = -\frac{1}{\alpha_i}(ln\lambda_1 - ln\beta_i), i \in M(g)$. Find the value of λ_1 by analogy with point 1,

$$\varphi_i = \frac{\Phi_g/\alpha_i}{\sum_{j \in M(g)} \frac{c_j}{\alpha_j}} - \frac{1}{\alpha_i} \frac{\sum_{j \in M(g)} \frac{c_j}{\alpha_j} ln\beta_j}{\sum_{j \in M(g)} \frac{c_j}{\alpha_j}} + \frac{ln\beta_i}{\alpha_i}$$

$p_i = \frac{c_i/\alpha_i}{\sum_{j \in M(g)} c_j/\alpha_j}$, where p_i is greater than zero. We find that if $\lambda_1 - \beta_j \geq 0$, then $\varphi_j = 0$. If $\lambda_1 - \beta_j < 0$, then it follows from point 1 that $\varphi_j =$

$-\frac{1}{\alpha_j}\left(ln\lambda_1 - ln\beta_j\right) > 0$, hence $j \in M(g)$. The final form of the solution is

$$\varphi_i = \left[\frac{\Phi_g/\alpha_i}{\sum_{j \in M(g)}\frac{c_j}{\alpha_j}} - \frac{1}{\alpha_i}\frac{\sum_{j \in M(g)}\frac{c_j}{\alpha_j}ln\beta_j}{\sum_{j \in M(g)}\frac{c_j}{\alpha_j}} + \frac{ln\beta_i}{\alpha_i}\right]^+,$$

$$p_i = \begin{cases} \frac{c_i/\alpha_i}{\sum_{j \in M(g)}c_j/\alpha_j}, & i \in M(g); \\ 0, & i \notin M(g). \end{cases}$$

$\square$

The challenge is that one needs to know in advance which values of $\varphi_i > 0$ are positive. The problem is solved just by searching through the values of φ_i.

There is a peculiarity to be observed in the solution. If the searcher allocates a resource of $\varphi_i > 0$ to vertex i, then the mobile objects also choose this vertex with non-zero probability. It would be logical for the mobile objects to move to the vertices to which no searching resource is allocated. The targets, however, do not move to such vertices, since the detection probability at the next stage would be rather high.

Example 3 Let the graph G be shown in Fig. 3, c_i, α_i parameter values as set in Table 3.

At the first time instant both mobile objects are at vertex zero. Imagine a situation where the searcher is ordered to allocate at least one unit resource to the putative position of the mobile objects (i.e., $\varphi_i \geq 1, i = 1, 2, \ldots, 9$). Set $\Phi = 100$. How should the searcher distribute the budget to maximize the probability of detecting the mobile players? We first need to find the optimal transition probabilities. Since it is stipulated than a non-zero resource is allocated to each vertex, then according to Theorem 3 the values of $p_i, i = 1, 2, \ldots, 9$ have the form

$$p_1 = \frac{c_1/\alpha_1}{\frac{c_1}{\alpha_1} + \frac{c_2}{\alpha_2} + \frac{c_3}{\alpha_3}} = 0, 24; \quad p_2 = \frac{c_2/\alpha_2}{\frac{c_1}{\alpha_1} + \frac{c_2}{\alpha_2} + \frac{c_3}{\alpha_3}} = 0, 12;$$

Fig. 3 Graph G

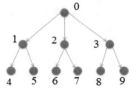

Table 3 Values of parameters c_i, α_i

i	1	2	3	4	5	6	7	8	9
α_i	0.9	0.8	0.3	0.5	0.4	0.7	0.2	0.8	0.9
c_i	9	4	8	5	8	7	9	6	3

$$p_3 = \frac{c_3/\alpha_3}{\frac{c_1}{\alpha_1} + \frac{c_2}{\alpha_2} + \frac{c_3}{\alpha_3}} = 0,64;$$

$$p_4 = \frac{c_4/\alpha_4}{\frac{c_4}{\alpha_4} + \frac{c_5}{\alpha_5}} = \frac{1}{3}; \; p_5 = \frac{c_5/\alpha_5}{\frac{c_4}{\alpha_4} + \frac{c_5}{\alpha_5}} = \frac{2}{3}; \; p_6 = \frac{c_6/\alpha_6}{\frac{c_6}{\alpha_6} + \frac{c_7}{\alpha_7}} = \frac{2}{11};$$

$$p_7 = \frac{c_7/\alpha_7}{\frac{c_6}{\alpha_6} + \frac{c_7}{\alpha_7}} = \frac{9}{11}; \; p_8 = \frac{c_8/\alpha_8}{\frac{c_8}{\alpha_8} + \frac{c_9}{\alpha_9}} = \frac{9}{13}; \; p_9 = \frac{c_9/\alpha_9}{\frac{c_8}{\alpha_8} + \frac{c_9}{\alpha_9}} = \frac{4}{13};$$

$$\varphi_4 = \frac{\Phi_1/\alpha_4}{\frac{c_4}{\alpha_4} + \frac{c_5}{\alpha_5}} = \frac{\Phi_1}{15}, \; \varphi_5 = \frac{\Phi_1/\alpha_5}{\frac{c_4}{\alpha_4} + \frac{c_5}{\alpha_5}} = \frac{\Phi_1}{12}, \; \varphi_6 = \frac{\Phi_2/\alpha_6}{\frac{c_6}{\alpha_6} + \frac{c_7}{\alpha_7}} = \frac{2\Phi_2}{77},$$

$$\varphi_7 = \frac{\Phi_2/\alpha_7}{\frac{c_6}{\alpha_6} + \frac{c_7}{\alpha_7}} = \frac{\Phi_2}{11}, \; \varphi_8 = \frac{\Phi_3/\alpha_8}{\frac{c_8}{\alpha_8} + \frac{c_9}{\alpha_9}} = \frac{3\Phi_2}{26}, \; \varphi_9 = \frac{\Phi_3/\alpha_9}{\frac{c_8}{\alpha_8} + \frac{c_9}{\alpha_9}} = \frac{4\Phi_2}{39}.$$

The values of β_1, β_2, β_3 will have the form

$$\beta_1 = exp\left(-\frac{\Phi_1}{\frac{c_4}{\alpha_4} + \frac{c_5}{\alpha_5}}\right), \; \beta_2 = exp\left(-\frac{\Phi_2}{\frac{c_6}{\alpha_6} + \frac{c_7}{\alpha_7}}\right), \; \beta_3 = exp\left(-\frac{\Phi_3}{\frac{c_8}{\alpha_8} + \frac{c_9}{\alpha_9}}\right).$$

The values of the variables φ_1, φ_2, φ_3 are calculated by the formulas

$$\varphi_1 = \frac{\Phi_0/\alpha_1}{\frac{c_1}{\alpha_1} + \frac{c_2}{\alpha_2} + \frac{c_3}{\alpha_3}} - \frac{1}{\alpha_1} \cdot \frac{\frac{c_1}{\alpha_1}\left(-\frac{\Phi_1}{\frac{c_4}{\alpha_4} + \frac{c_5}{\alpha_5}}\right) + \frac{c_2}{\alpha_2}\left(-\frac{\Phi_2}{\frac{c_6}{\alpha_6} + \frac{c_7}{\alpha_7}}\right) + \frac{c_3}{\alpha_3}\left(-\frac{\Phi_3}{\frac{c_8}{\alpha_8} + \frac{c_9}{\alpha_9}}\right)}{\frac{c_1}{\alpha_1} + \frac{c_2}{\alpha_2} + \frac{c_3}{\alpha_3}} +$$

$$+ \frac{1}{\alpha_1}\left(-\frac{\Phi_1}{\frac{c_4}{\alpha_4} + \frac{c_5}{\alpha_5}}\right)$$

$$\varphi_2 = \frac{\Phi_0/\alpha_2}{\frac{c_1}{\alpha_1} + \frac{c_2}{\alpha_2} + \frac{c_3}{\alpha_3}} - \frac{1}{\alpha_2} \cdot \frac{\frac{c_1}{\alpha_1}\left(-\frac{\Phi_1}{\frac{c_4}{\alpha_4} + \frac{c_5}{\alpha_5}}\right) + \frac{c_2}{\alpha_2}\left(-\frac{\Phi_2}{\frac{c_6}{\alpha_6} + \frac{c_7}{\alpha_7}}\right) + \frac{c_3}{\alpha_3}\left(-\frac{\Phi_3}{\frac{c_8}{\alpha_8} + \frac{c_9}{\alpha_9}}\right)}{\frac{c_1}{\alpha_1} + \frac{c_2}{\alpha_2} + \frac{c_3}{\alpha_3}} +$$

$$+ \frac{1}{\alpha_2}\left(-\frac{\Phi_2}{\frac{c_6}{\alpha_6} + \frac{c_7}{\alpha_7}}\right)$$

$$\varphi_3 = \frac{\Phi_0/\alpha_3}{\frac{c_1}{\alpha_1} + \frac{c_2}{\alpha_2} + \frac{c_3}{\alpha_3}} - \frac{1}{\alpha_3} \cdot \frac{\frac{c_1}{\alpha_1}\left(-\frac{\Phi_1}{\frac{c_4}{\alpha_4} + \frac{c_5}{\alpha_5}}\right) + \frac{c_2}{\alpha_2}\left(-\frac{\Phi_2}{\frac{c_6}{\alpha_6} + \frac{c_7}{\alpha_7}}\right) + \frac{c_3}{\alpha_3}\left(-\frac{\Phi_3}{\frac{c_8}{\alpha_8} + \frac{c_9}{\alpha_9}}\right)}{\frac{c_1}{\alpha_1} + \frac{c_2}{\alpha_2} + \frac{c_3}{\alpha_3}} +$$

$$+ \frac{1}{\alpha_3} \left(-\frac{\Phi_3}{\frac{c_8}{\alpha_8} + \frac{c_9}{\alpha_9}} \right).$$

One can also check that $c_1\varphi_1 + c_2\varphi_2 + c_3\varphi_3 = \Phi_0$.. Simplifying the values $\varphi_1, \varphi_2, \varphi_3$, we get $\varphi_1 = \frac{2\Phi_0}{75} - \frac{19\Phi_1}{675} + \frac{2\Phi_2}{825} + \frac{64\Phi_3}{975}$, $\varphi_2 = \frac{3\Phi_0}{100} + \frac{\Phi_1}{100} - \frac{\Phi_2}{50} + \frac{24\Phi_3}{325}$, $\varphi_3 = \frac{2\Phi_0}{25} + \frac{2\Phi_1}{75} + \frac{2\Phi_2}{275} - \frac{36\Phi_3}{325}$. Since $\beta_i e^{-\alpha_i \varphi_i} = \lambda_1$, the probability of detecting at least one mobile player is

$$1 - \sum_{i \in ch(g)} \beta_i^2 (p_i^*)^2 e^{-2\alpha_i \varphi_i} - \sum_{i \in ch(g)} \sum_{j \in ch(g), i \neq j} \beta_i \beta_j p_i^* p_j^* e^{-\alpha_i \varphi_i - \alpha_j \varphi_j} =$$

$$= 1 - \sum_{i \in ch(g)} (p_i^*)^2 \lambda_1^2 - \sum_{i \in ch(g)} \sum_{j \in ch(g), i \neq j} p_i^* p_j^* \lambda_1^2 = 1 - \lambda_1^2,$$

where $\lambda_1 = exp\left(\frac{-\Phi_0 + \sum_{j=1,2,3} \frac{c_j}{\alpha_j} ln\beta_j}{\sum_{j=1,2,3} \frac{c_j}{\alpha_j}} \right)$. The searcher seeks to distribute the budget so as to maximize the detection probability, i.e. the magnitude $1 - \lambda_1^2$. To achieve this it suffices to minimize $-\Phi_0 + \sum_{j=1,2,3} \frac{c_j}{\alpha_j} ln\beta_j = -\Phi_0 - \frac{\Phi_1}{3} - \frac{\Phi_2}{11} - \frac{32\Phi_3}{13}$. To find $\Phi_0, \ldots, \Phi_3$ it is enough to solve the linear programming problem

$$-\Phi_0 - \frac{\Phi_1}{3} - \frac{\Phi_2}{11} - \frac{32\Phi_3}{13} \to min$$

$$\begin{cases} \frac{2\Phi_0}{75} - \frac{19\Phi_1}{675} + \frac{2\Phi_2}{825} + \frac{64\Phi_3}{975} \geq 1, \\ \frac{3\Phi_0}{100} + \frac{\Phi_1}{100} - \frac{\Phi_2}{50} + \frac{24\Phi_3}{325} \geq 1, \\ \frac{2\Phi_0}{25} + \frac{2\Phi_1}{75} + \frac{2\Phi_2}{275} - \frac{36\Phi_3}{325} \geq 1 \\ \frac{\Phi_1}{15} \geq 1, \\ \frac{\Phi_1}{12} \geq 1, \\ \frac{2\Phi_2}{77} \geq 1, \\ \frac{\Phi_2}{11} \geq 1, \\ \frac{3\Phi_3}{26} \geq 1, \\ \frac{4\Phi_3}{39} \geq 1, \\ \Phi_0 + \Phi_1 + \Phi_2 + \Phi_3 = 100. \end{cases}$$

Its solution results in $\Phi_0 = 28,677$; $\Phi_1 = 15,000$; $\Phi_2 = 38,500$; $\Phi_3 = 17,823$.

8 Solution for a Multi-stage Game

Having solved the single-stage game, one can proceed to a game with two stages. Knowing the solution of the two-stage game, one can get the solution for the three-stage game and so forth by mathematical induction. At each stage the payoff function retains its form, enabling the solution of the multi-stage game.

Theorem 5 *Let both mobile objects be at vertex g at stage n. Then, the optimal strategies for the players have the form*

$$\varphi_i^* = \left[\frac{\Phi_g/\alpha_i}{\sum_{j\in M(j)} \frac{c_j}{\alpha_j}} - \frac{1}{\alpha_i} \frac{\sum_{j\in M(g)} \frac{c_j}{\alpha_j} ln\beta_j}{\sum_{j\in M(g)} \frac{c_j}{\alpha_j}} + \frac{ln\beta_i}{\alpha_i} \right]^+ ,$$

$$p_i^* = \begin{cases} \frac{c_i/\alpha_i}{\sum_{j\in M(g)} c_j/\alpha_j}, & i \in M(g); \\ 0, & i \notin M(g), \end{cases}$$

where $M(g)$—is the offspring set of vertex g, for which $\varphi_i > 0$.

Proof Note that the optimal probabilities of detecting players in a two-stage game can be transformed to the form $H_1^*(2, g, \Phi_g) = H_2^*(2, g, \Phi_g) = 1 - \beta_g$, $H_2^*(2, g, \Phi_g) = 1 - \beta_g^2$. A three-stage game is therefore solved analogously with a two-stage game. Applying the same reasoning to each stage, we are led by induction to the solution at stage n written in the form analogous to the solution at stage 2, as was to be proved. □

9 Solution for a Single-Stage Game with an Arbitrary Detection Probability

In most studies the detection probability of a target was equaled to $1 - e^{-\alpha_i \varphi_i}$. If the amount of the resource is sent to infinity, such detection probability will tend rapidly to 1. In practice, however, where the resource amounts are great, the methods and algorithms of tracking a mobile object may not be as simple, i.e. other functions setting the capture probability should be considered instead of the exponential detection probability.

Let $f(\alpha_i \varphi_i)$ be the probability of not detecting an object at vertex i. The higher the φ_i, the lower the probability of not detecting the object. We therefore believe that $f(\alpha_i \varphi_i)$ is a strictly decreasing function. If the resource amount allocated to vertex i equals zero, the non-detection probability equals 1, i.e. $f(0) = 1$. The property of probability dictates that the inequality $f(\alpha_i \varphi_i) > 0$ must hold. The value of the function $f(\alpha_i \varphi_i) \neq 0 \forall \varphi_i$ because $f(\alpha_i \varphi_i)$ is strictly decreasing.

Let $H(P, \Psi) = \sum_{i=1}^{n} p_i f(\alpha_i \varphi_i)$, $p_i \geq 0$, $\varphi_i \geq 0$, $p_1 + p_2 + \ldots + p_n = 1$, $c_1 \varphi_1 + \ldots + c_n \varphi_n \leq \Phi$

Theorem 6 *Let $f(x)$ be a differentiable on the interval $[0; +\infty)$, convex downward, strictly decreasing function. Then, for the zero-sum game $\Gamma = \langle I, II; P, \Psi; H(P, \Psi) \rangle$ the value of the game and the equilibrium point have the form*

$$H^*(P^*, \Psi^*) = f\left(\frac{\Phi}{\sum_{i=1}^{n} \frac{c_i}{\alpha_i}}\right), \; p_i^* = \frac{c_i/\alpha_i}{\sum_{i=1}^{n} \frac{c_i}{\alpha_i}}, \; \varphi_i^* = \frac{\Phi/\alpha_i}{\sum_{i=1}^{n} \frac{c_i}{\alpha_i}}.$$

Proof To find the equilibrium point we write down the required Kuhn–Tacker conditions.

$$\begin{cases} -f(\alpha_i \varphi_i) - U_i + \lambda_1 = 0; \\ \alpha_i p_i f'(\alpha_i \varphi_i) - W_i + c_i \lambda_2 = 0; \\ U_i p_i = 0; \\ W_i \varphi_i = 0; \\ p_1 + \ldots + p_n = 1; \\ c_1 \varphi_1 + \ldots + c_n \varphi_n = \Phi; \\ p_i, \varphi_i, u_i, W_i \geq 0 \end{cases}$$

where $i = 1, 2, \ldots, n$. Note that there is a minus in front of $f(\alpha_i \varphi_i)$ in the first n equations of the system, since the maximum is found from the variables p_i.

If $\forall i : U_i = W_i = 0$, то $f(\alpha_i \varphi_i^*) = \lambda_1$. Since $f(x)$ is a strictly decreasing function, there exists an inverse function $f^{-1}(y)$. Then $\alpha_i \varphi_i^* = f^{-1}(\lambda_1)$, $\varphi_i^* = \frac{1}{\alpha_i} f^{-1}(\lambda_1)$. Since $\sum_{i=1}^{n} c_i \varphi_i^* = \Phi$, то $f^{-1}(\lambda_1) = \frac{\Phi}{\sum_{i=1}^{n} \frac{c_i}{\alpha_i}}$, $f(f^{-1}(\lambda_1)) = f\left(\frac{\Phi}{\sum_{i=1}^{n} \frac{c_i}{\alpha_i}}\right)$, $\lambda_1 = f\left(\frac{\Phi}{\sum_{i=1}^{n} \frac{c_i}{\alpha_i}}\right)$. Given the equality $f(\alpha_i \varphi_i^*) = \lambda_1$, we get

$H^*(P^*, \Psi^*) = \sum_{i=1}^{n} p_i^* f(\alpha_i \varphi_i^*) = \lambda_1 \sum_{i=1}^{n} p_i^* = \lambda_1 = f\left(\frac{\Phi}{\sum_{i=1}^{n} \frac{c_i}{\alpha_i}}\right)$. Value

$\varphi_i^* = \frac{1}{\alpha_i} f^{-1}(\lambda_1) = \frac{\Phi/\alpha_i}{\sum_{j=1}^{n} \frac{c_j}{\alpha_j}}$.

Find p_i^*. $\alpha_i p_i^* f'(\alpha_i \varphi_i^*) = -c_i \lambda_2$, $\alpha_i p_i^* f'\left(\alpha_i \cdot \frac{\Phi/\alpha_i}{\sum_{j=1}^{n} \frac{c_j}{\alpha_j}}\right) = -c_i \lambda_2$, $f'\left(\frac{\Phi}{\sum_{j=1}^{n} \frac{c_j}{\alpha_j}}\right)$

$= -\frac{c_i}{\alpha_i p_i^*} \cdot \lambda_2$. The expression on the left in the last equality is independent of i, hence $\frac{c_1}{\alpha_1 p_1^*} \cdot \lambda_2 = \frac{c_2}{\alpha_2 p_2^*} \cdot \lambda_2 = \ldots = \frac{c_n}{\alpha_n p_n^*} \cdot \lambda_2$. Since $\sum_{i=1}^{n} p_i^* = 1$, we get $p_i^* = \frac{c_i/\alpha_i}{\sum_{j=1}^{n} \frac{c_j}{\alpha_j}}$. The resultant values of p_i^*, φ_i^*, and $U_i = W_i = 0 \; \forall i$ satisfy all the equations in the system. Since the optimized function is convex and restrictions on the variables are linear, the eventual solution represents an equilibrium point. $\square$

Acknowledgements This work was supported by the Russian Fund for Basic Research (projects 16-01-00183 and 16-51-55006).

References

1. Alpern, Steve. "Hide-and-Seek Games on a Network, Using Combinatorial Search Paths." Operations Research 65.5 (2017): 1207–1214.
2. Baston, Vic, and Kensaku Kikuta. "Search games on a broken wheel with traveling and search costs" Journal of the Operations Research Society of Japan 60.3 (2017): 379–392.
3. Gal S (2005) Strategies for searching graphs. Golumbic M, Hartman I, eds. Graph Theory, Combinatorics and Algorithms, Vol. 34, Operations Research/Computer Science Interfaces Series (Springer, New York), 189–214.
4. Kekka, Toshiyuki, and Ryusuke Hohzaki. "A nonlinear model of a search allocation game with false contacts." Scienticae Mathematicae Japonicae 76 (2013): 497–515.
5. Hohzaki, Ryusuke. "A cooperative game in search theory." Naval Research Logistics (NRL) 56.3 (2009): 264–278.
6. Hohzaki, Ryusuke. "A search game with incomplete information on detective capability of searcher." Contributions to Game Theory and Management 10.0 (2017): 129–142.
7. Hohzaki, Ryusuke. "Search games: Literature and survey." Journal of the Operations Research Society of Japan 59.1 (2016): 1–34.
8. Hohzaki, Ryusuke. "A multi-stage search allocation game with the payoff of detection probability." Journal of the Operations Research Society of Japan 50.3 (2007): 178–200.

The Impact of Product Differentiation on Symmetric R&D Networks

Mohamad Alghamdi, Stuart McDonald, and Bernard Pailthorpe

Abstract This paper examines the impact of product differentiation on an R&D network. We find that when firms produce goods that are complements or independent, R&D expenditure, prices, firms? net profits and total welfare are always higher under price competition than under quantity competition. When goods are substitutes, R&D expenditure and profits are higher under quantity competition than under price competition. Also, when goods are substitutes and products are sufficiently differentiated, then total welfare is higher in the Bertrand equilibrium than under the Cournot equilibrium. Beyond this threshold level of product differentiation, Cournot competition is superior in terms of social welfare. The paper finds that the key threshold level of product differentiation, determining the relative superiority of the Cournot and Bertrand equilibrium when goods are substitutes, depends on the cost efficiency of R&D and the number of collaborative partnerships that firms participate in relative to the size of the network. We show that when goods are substitutes, if the network is dense so that the number of partnerships is large relative to the number of firms operating in the market, then the threshold value of the product differentiation parameter can be small.

Keywords Research and Development (R&D) · Symmetric networks · Cooperative links · Oligopoly · Product differentiation

JEL Classification D21; D43

M. Alghamdi
Mathematics, King Saud University, Riyadh, Saudi Arabia
e-mail: almohamad@ksu.edu.sa

S. McDonald (✉)
School of Economics, University of Nottingham, Ningbo, China
e-mail: Stuart.McDonald@nottingham.edu.cn

B. Pailthorpe
School of Physics, University of Sydney, Sydney, NSW, Australia
e-mail: Bernard.Pailthorpe@Sydney.edu.au

© Springer Nature Switzerland AG 2020
D. Yeung et al. (eds.), *Frontiers in Games and Dynamic Games*, Annals of the
International Society of Dynamic Games 16,
https://doi.org/10.1007/978-3-030-39789-0_6

175

1 Introduction

There is substantial literature in economics that concentrates on the effect of imperfect appropriability of knowledge acquired from the innovation process (spillover) on the incentive to innovate (e.g., [2, 5, 6, 15, 18, 21, 23, 24]). This non-tournament R&D literature has tried to provide an answer to what extent research alliance, such as research joint ventures (RJVs) and research cartels, can facilitate knowledge transfer and cost sharing so as to promote R&D activity in a way that is welfare enhancing. However, in this literature spillovers are typically treated as being symmetric, impacting on all firms to the same degree. When asymmetries are introduced, the models tend to involve only a small number of firms, with very specific assumptions regarding the costs and productivity of R&D. As a consequence, this literature on research consortia does not provide an adequate answer to the problem of quantifying the direction and extent to which R&D activity spills over to outside firms.

One approach that can address this important question is to combine concepts from network theory with the economic theory of R&D, so as to explicitly model R&D alliances by using a network (e.g., [7–9, 17, 22, 25, 26]). Within this class models, this paper examines the relationship between cooperative R&D agreements and market structure in an oligopolistic setting, with linear demand and product differentiation. As in [9] firms strategically form bi-lateral collaborative links with competing firms, for the purpose of sharing knowledge generated from the R&D process. The aim of this chapter is to examine the effects of three factors: structure of the market, cooperative links and knowledge spillover on R&D investment and profit of firms, and their effects on strategic stability of R&D agreements and on welfare.

This paper shows that there are qualitative differences between Cournot and Bertrand competition in terms of R&D effort, profit, and total welfare. These differences depend on the differentiation degree and number of cooperative links. If products are complements, Bertrand competition dominates Cournot competition from the perspective of the firm, since profits are always higher. However, if goods are substitutes, firms' profits are always higher under Cournot competition. This is because R&D effort is higher in Bertrand competition if goods are complements, but lower if goods are substitutes. In terms of welfare, price is lower and production is always higher under Bertrand competition than under Cournot, regardless of number of firms and type of product (complement or substitute), and structure of the network. Moreover, when goods are complements, it is found that Bertrand competition is socially preferable because total welfare is higher than under Cournot.

Our result confirms that consumer surplus is always higher under Bertrand competition, which is consistent with [12]. However, total welfare is higher under Bertrand competition except if goods are substitutes and the network is dense, in the sense that the majority of firms are connected. This result stands apart from the conventional wisdom that Bertrand competition delivers higher welfare [20]. This result extends prior works by Singh and Vives [20], Hackner [10], Hsu and Wang [12] and indicates that the structure of the network is a matter in the comparison between Cournot and Bertrand in terms of the total welfare.

Within the non-tournament R&D literature the evidence regarding the welfare effects of R&D alliances are quite mixed. For example, under Cournot competition, it is known that R&D cooperation raises both equilibrium R&D investments and social welfare, when the spillover rate in R&D is high [16, 23]. However, [24] has shown that when the spillover rate is not high predictions regarding the welfare effects of R&D cooperation are not as conclusive. He shows that cooperative R&D reduces both equilibrium R&D and social welfare for intermediate spillovers and that cooperative R&D reduces the investments but has ambiguous effects on social welfare for low spillovers. More recently [11] have provided a comprehensive study on the impact of product differentiation on the non-tournament R&D model, complementing the study in [8] for R&D networks. Our paper extends their results, showing that when goods are substitutes, there is a threshold effect in the presence of R&D networks that determines the relative welfare optimality of Bertrand versus Cournot competition.

2 The Model

This paper focuses on an oligopolistic market in which firms produce a horizontally differentiated good using an ex ante identical production technology. Prior to competing in the product market, each firm has the opportunity to invest in the R&D of a cost-reducing technology. Firms also have the opportunity to choose to share this R&D by forming a bi-lateral agreements with other firms. As in [9], these bi-lateral agreements can encompass loose and informal agreements, such as memorandum of understanding, as well as more formal agreements specifying the sharing of research costs and ownership of intellectual property, such as research joint ventures. These bi-lateral R&D agreements between firms define a network of collaboration—an R&D network—in which the firms constitute the network nodes and the bi-lateral R&D constitute the edges of the network. The sequence of play between competing firms in this model, which leads to the formation of this R&D network, will follow that of [9]:

The First Stage Each firm chooses its research partners. Firms collaborate by forming bi-lateral (or pairwise) links between themselves and other firms. The firms and the cooperative links together characterize a network of cooperation in R&D.

The Second Stage Given the R&D network, each firm chooses the amounts of investment (efforts) in R&D simultaneously and independently in order to reduce the cost of production. The R&D effort across the network determines the effective R&D of each firm.

The Third Stage Given the R&D investments of each firm and the effective R&D effort (as determined by the R&D network), firms now compete in the product market by setting quantities (Cournot competition) or prices (Bertrand competition) in order to maximize their profits.

The structure of the individual components of this model is now defined formally below.

R&D Networks Throughout this paper an R&D network G is defined as an undirected graph (N, E) composed of a set of nodes $N(G) = \{1, 2, \ldots, n\}$ representing the population of firms operating in this market, and a set of edges or links $E(G)$, where each link ij in this set represents a RJV agreement between firms i and $j \in N$. If $|N| = n$ is the number of nodes and $|E| = m$ is the number of links, the density of the network G is $D = 2m/n(n-1)$. The set $N_i = \{j \in N; ij \in E(G)\}$ is the neighborhood set of firm i and is the set of firms participating in RJV agreements with firm i [13]. The number of firms participating in RJV agreements with firm i is given by $k_i = |N_i|$, the degree of node i (i.e., the number of edges incident to node i). The maximum and minimum degree are given by $k_{max}(G)$ and $k_{min}(G)$, respectively, and denote the highest and lowest number of collaborations by any firm in the R&D network.

This paper focuses on symmetric R&D networks, in which each firm forms an identical number of R&D agreements. Formally, a symmetric R&D network of size k, G_k, is an undirected graph (N, E) in which each node (firm) has the same degree k (i.e., for each $i \in N$, $N_i = k$). In the network theory literature the R&D network G_k is called the k-regular network, where k denotes the common degree size. The complete network, K_n, is a graph such that for any pair of firms $i, j \in N$, there exists a RJV agreement linking them. In our notation $G_{n-1} = K_n$, i.e., complete network is a symmetric R&D network in which each firm (node) has $n-1$ connections. The empty network, E_n, is composed of firms investing in R&D, but not co-operating; therefore, $E_n = G_0$.

Demand Side There are n differentiated products. The utility function of consumers is a generalization of the quadratic utility function given by Singh and Vives [20] and Hackner [10],

$$U(q_1, q_2, \ldots, q_n) = a \sum_{i=1}^{n} q_i - \frac{1}{2}\Big(\alpha \sum_{i=1}^{n} q_i^2 + 2\lambda \sum_{j \neq i} q_i q_j\Big) + m, \quad -1 \leq \lambda \leq 1.$$

(1)

Here the demand parameter $a > 0$ denotes the willingness of consumers to pay and $a > 0$ is the diminishing marginal rate of consumption, while q_i is the quantity consumed of good i and m measures the consumer's consumption of all other products. Without loss of generality, it is assumed that $\alpha = 1$ to simplify the analysis. The utility function is concave function, so the first order condition determines the optimal consumption for good i. The inverse demand function for each good i is given by

$$p_i = D_i^{-1}(q_i, q_2, \ldots, q_n) = a - q_i - \lambda \sum_{j \neq i} q_j, \quad i = 1, 2, \ldots, n, \tag{2}$$

where the domain of λ is as denoted in (1). The demand equation for good i is

$$q_i = D_i(p_i, p_2, \ldots, p_n) = \frac{(1 - \lambda)a - (1 + (n - 2)\lambda)p_i + \lambda \sum_{j \neq i} p_j}{(1 - \lambda)(1 + (n - 1)\lambda)},$$

$$i = 1, 2, \ldots, n. \tag{3}$$

Goods are substitutes if $\lambda > 0$, if $\lambda < 0$ they are complements and they are independents if $\lambda = 0$. If $\lambda = 1$, the goods are perfect substitutes and if $\lambda = -1$, the goods are perfect complements.[1]

Supply Side The supply side of this market is composed of n oligopolistic firms, where each firm in a market produces a single differentiated product. The profit of the i'th firm is defined by the following expression:

$$\pi_i = (p_i - c_i)D_i(p_i, p_2, \ldots, p_n) - \gamma x_i^2, \quad i = 1, 2, \ldots, n, \tag{4}$$

where c_i is the marginal cost, x_i is the R&D investment of the i'th firm, and γx_i^2 is the cost of investment in R&D, with $\gamma > 0$ measures the rate of change in the cost of R&D. We follow [9] and characterize firm i's marginal cost by

$$c_i = \max \left\{ 0, \bar{c} - x_i - \sum_{j \in N_i} x_j - \mu \sum_{k \neq N_i} x_k \right\}, \quad i = 1, 2, \ldots, n, \tag{5}$$

where $\bar{c}$ is the marginal cost of production, x_i, x_j and x_k, $k, j \neq i$, denote the cost-reducing R&D investment of firm i, its partners, and its competitors, N_i is the set of firms participating in a joint venture with firm i, and $\mu \in [0, 1)$ is an exogenous parameter that captures knowledge spillovers acquired from firms not engaged in a joint venture with firm i. Since the quasi-linear utility function (1) is measured in money, total welfare must be given by U (as defined by (1)) minus the sum of production and R&D costs:

$$TW = U(q_1, q_2, \ldots, q_n) - \sum_{i=1}^{n} \left(c_i q_i - \gamma x_i^2 \right). \tag{6}$$

---

[1]The differentiation degree also measures the competition rate between firms. When goods are complements or independent, firms are in a weakly competitive market. Whereas, if goods are substitutes or homogeneous, firms are in a competitive market where increase $\lambda \to 1$ leads to increase the competition intensity between firms.

Stability and Efficiency of R&D Networks The study of R&D cooperation under the network game involves the concepts of pairwise stability and efficiency. The pairwise stability depends on firms' profit functions and it is a necessary condition for strategic stability as shown in [14]. The definition of the efficiency of a network is given as follows and is determined by the total welfare generated from that network.

Definition 1 (Pairwise Stability) For any network G to be stable, the following two conditions need to be satisfied for any two firms $i, j \in G$:

1. If $ij \in G$, $\pi_i(G) \geq \pi_i(G - ij)$ and $\pi_j(G) \geq \pi_j(G - ij)$,
2. If $ij \notin G$ and if $\pi_i(G) < \pi_i(G + ij)$, then $\pi_j(G) > \pi_j(G + ij)$, where $\pi_i(\cdot)$ and $\pi_j(\cdot)$ denote the profits firm i and firm j as defined by Eq. (4).

$G - ij$ is the network resulting from deleting a link ij from the network G and $G + ij$ is the network resulting from adding a link ij to the network G.

Definition 2 (Network Efficiency) Network G is said to be efficient if no other network $\acute{G}$ can be generated by adding or deleting links, such that $TW(\acute{G}) > TW(G)$, where $TW(\cdot)$ denotes total welfare as defined in Eq. (6).

Strategic Complement and Substitute Players' strategies are said to be strategic complements (substitutes) if the best response of player i to the choice of player j is increasing (decreasing) in the choice parameter of j [3]. Amir and Jin [1] note that under linear demand, in Cournot competition, production quantities are called strategic complements if firm j raises its output and the other firms increase their prices. This is translated into the expression $\frac{\partial p_i}{\partial q_j} \geq 0$ for all goods $i \neq j$. In Bertrand competition, prices of the products are called strategic complements when firm j raises its price, the other firms increase their outputs. In other words, $\frac{\partial q_i}{\partial p_j} \geq 0$ for all $i \neq j$.

Under Cournot competition, from the inverse demand function, $\frac{\partial p_i}{\partial q_j} = -\lambda$. Then, if $\lambda > 0$, $\frac{\partial p_i}{\partial q_j} < 0$ which means if goods are substitutes, the quantities are strategic substitutes. Whereas if $\lambda \leq 0$, $\frac{\partial p_i}{\partial q_j} \geq 0$ which indicates that if goods are complements or independent, the quantities are strategic complement.

Under Bertrand competition,

$$q_i = \frac{(1 - \lambda)a - (1 + (n - 2)\lambda) p_i + \lambda \sum_{j \neq i} p_j}{(1 - \lambda)(1 + (n-1)\lambda)} \Rightarrow \frac{\partial q_i}{\partial p_j} = \frac{\lambda}{(1-\lambda)(1 + (n-1)\lambda)}.$$

Then, if $\lambda \geq 0$, $\frac{\partial q_i}{\partial p_j} \geq 0$. This indicates that if products are substitutes or independent, then the prices are strategic complement. However, if $\frac{1}{1-n} < \lambda < 0$ (see Proposition 4), then $\frac{\partial q_i}{\partial p_j} < 0$. This means that if products are complementary, then the prices are strategic substitutes.

3   Cournot Competition

This section studies the impact of product differentiation on the R&D network under Cournot competition. Substituting the inverse demand function, Eq. (2) into Eq. (4) gives the profit function for firm i

$$\pi_i = \left(a - q_i - \lambda \sum_{j \neq i} q_j\right) q_i - c_i q_i - \gamma x_i^2, \quad i = 1, \ldots, n . \tag{7}$$

The best response function for firm i, $q_i = (a - c_i - \lambda \sum_{j \neq i} q_j)/2$, $i = 1, 2, \ldots, n$, can be derived from the first order maximizing condition for Eq. (7). The Nash equilibrium output for each firm i in the product market stage game is then derived by solving the resulting system of best response functions:

$$q_i^* = \frac{(2 - \lambda)a - (2 + (n - 2)\lambda)c_i + \lambda \sum_{j \neq i} c_j}{(2 - \lambda)\big((n - 1)\lambda + 2\big)} . \tag{8}$$

To find the Nash equilibrium profit, the equilibrium output (8) is substituted into the profit function which gives

$$\pi_i^* = \left[\frac{(2 - \lambda)a - (2 + (n - 2)\lambda)c_i + \lambda \sum_{j \neq i} c_j}{(2 - \lambda)\big((n - 1)\lambda + 2\big)}\right]^2 . \tag{9}$$

Let the effort of firm i in the R&D network G_k be x_i and other firms that linked to firm i with subscript r invest x_r in R&D. Also, the remaining firms $n - k - 1$, that are not linked to firm i, are represented with subscript p and invest x_p in R&D. Thus, there are three structures for the production cost function defined in Eq. (5):

$$c_i = \bar{c} - x_i - k x_r, \quad c_r = \bar{c} - x_r - \sum_{l \in N_r} x_l, \quad c_p = \bar{c} - x_p - \sum_{l \in N_p} x_l . \tag{10}$$

Substituting these cost functions into the profit function (9) results in the expression of the profit of firm i as a function of the R&D investment decisions of itself, its k collaborators, and other $n - k - 1$ firms contained in the network:

$$\pi_i =$$

$$\frac{\left[(2-\lambda)(a-\bar{c}) + (2 + (n - (k+2))\lambda)x_i + k((n - (k+2))\lambda+2)x_r - \lambda(k + 1)(n-k-1)x_p\right]^2}{\left((2-\lambda)((n - 1)\lambda + 2)\right)^2} - \gamma x_i^2 . \tag{11}$$

The Nash equilibrium of the R&D stage game can be derived by solving the resulting system first order conditions derived from the profit maximizing problem of each firm i and results in a symmetric Nash equilibrium. The Nash equilibrium for the R&D stage game is expressed by the following formula:

R&D Effort

$$x^* = \frac{\big((n-(k+2))\lambda + 2\big)(a - \overline{c})}{\gamma\big((n-1)\lambda + 2\big)^2(2-\lambda) - (k+1)\big((n-(k+2))\lambda + 2\big)}. \tag{12}$$

By substituting this effort function into the cost, quantity, profit, and total welfare functions, the following equilibria are obtained:

Cost Function

$$c^* = \frac{\overline{c}\gamma\big((n-1)\lambda + 2\big)^2(2-\lambda) - (k+1)\big((n-(k+2))\lambda + 2\big)a}{\gamma\big((n-1)\lambda + 2\big)^2(2-\lambda) - (k+1)\big((n-(k+2))\lambda + 2\big)}, \tag{13}$$

Quantity Function

$$q^* = \frac{\gamma(2-\lambda)\big((n-1)\lambda + 2\big)(a - \overline{c})}{\gamma\big((n-1)\lambda + 2\big)^2(2-\lambda) - (k+1)\big((n-(k+2))\lambda + 2\big)}, \tag{14}$$

Profit Function

$$\pi^* = \frac{\gamma\left[\gamma(2-\lambda)^2\big((n-1)\lambda + 2\big)^2 - \big((n-(k+2))\lambda + 2\big)^2\right](a-\overline{c})^2}{\left[\gamma\big((n-1)\lambda + 2\big)^2(2-\lambda) - (k+1)\big((n-(k+2))\lambda + 2\big)\right]^2}, \tag{15}$$

Total Welfare Function

$$TW^* = \frac{n\gamma\left[\gamma(2-\lambda)^2\big((n-1)\lambda + 2\big)^2\big(3+(n-1)\lambda\big) - 2\big((n-(k+2))\lambda + 2\big)^2\right](a-\overline{c})^2}{2\left[\gamma\big((n-1)\lambda + 2\big)^2(2-\lambda) - (k+1)\big((n-(k+2))\lambda + 2\big)\right]^2}. \tag{16}$$

The R&D efficiency parameter γ measures the cost effectiveness of firm-level R&D investment. Lemma 1 shows that to obtain a positive level of investment in R&D, which also satisfies the sufficient conditions of the Nash equilibrium, the R&D efficiency parameter γ must exceed a threshold that depends on the size and structure of the market, the level of cooperation between firms within the R&D network.

Lemma 1 (R&D Efficiency Parameter) *For Cournot competition with n firms participating in a k-regular R&D network and spillover parameter $\mu = 0$, R&D parameter γ must satisfy the following condition:*

$$\gamma > max \left\{ \frac{(k+1)\big((n-(k+2))\lambda+2\big)a}{\bar{c}\big((n-1)\lambda+2\big)^2(2-\lambda)}, \left[\frac{(n-(k+2))\lambda+2}{\big((n-1)\lambda+2\big)(2-\lambda)} \right]^2 \right\}.$$

(17)

The proof is given in the Appendix 1.

The next lemma concerns the impact that λ has on the Nash equilibrium output. The condition in the lemma implies that when firms produce complementary goods, the overall size of firms participating in the R&D network is restricted by the degree of complementarity between products. In particular, when firms produce perfect complements there can be no more than three firms in the R&D network.

Lemma 2 (Substitution Degree) *For Cournot competition with n firms participating in a k-regular R&D network and zero spillover, the substitution degree should satisfy the following condition:*

$$\lambda > \frac{2}{1-n}.$$

The proof is given in the Appendix 1.

The Proposition 1 provides a characterization of R&D investment under Cournot competition. It considers three cases: Case 1 shows that there is a negative relationship between R&D effort and the product differentiation degree λ. Cases 2 and 3 show the impact of the cooperative activity level k on R&D expenditure. Case 2 shows that if goods are complements or independent, then firm-level R&D expenditure will always increase as k increases. Case 3 shows that this relationship will hold when goods are substitutes if $\lambda \leq \bar{\lambda}$. However, if $\lambda > \bar{\lambda}$, firm-level R&D expenditure decreases as k increases. Case 2 can be regarded as an extension of Proposition 1 in [9], which focuses exclusively on the case of independent markets. However, Case 3 shows that product differentiation degree is important for determining the impact of the level of cooperation. When $\lambda > \bar{\lambda}$, the effect described in Proposition 2 of [9] occurs, where R&D investment declines as k increases. Moreover, the threshold level of λ depends not only on n and k, but also on γ, the parameter which governs the cost efficiency of R&D. When combined with Lemma 1, Case (3) of Proposition 1 shows that when goods are substitutes there is a non-linear relationship between λ, γ, n, and k. Most importantly, when $\lambda > 0$ positive levels of investment in R&D are sustainable only when firms participate in cooperative research agreements.

Proposition 1 (R&D Effort) *For Cournot competition with n firms participating in a k-regular R&D network,*

1. *The R&D investment of firms declines as the differentiation degree* λ *increases.*
2. *When outputs are strategic complements, then R&D investment is increasing as the cooperative activity level k increases in the R&D network.*
3. *When outputs are strategic substitutes, if the substitution degree* $\lambda \leq \bar{\lambda}$, *then R&D investment increases with k. For* $\lambda > \bar{\lambda}$, *R&D investment decreases with k—see Example 2 in Appendix 2.*

The proof is given in the Appendix 1.

Note that the lower bound $\bar{\lambda}$ for Case 3 of Proposition 1 depends on the size of the network n. Let G_{k_1} and G_{k_2} be symmetric networks with k_1 and k_2 links, respectively. Assume $k_1 > k_2$ and substitute k_1 and k_2 into the effort function (12). Then, by assuming $(a - \bar{c}) = 1$,

$$
x^*(G_{k_1}) - x^*(G_{k_2}) = \frac{(k_1 - k_2)\Big((n - (k_1 + 2))\lambda + 2\Big)}{\gamma\big((n - 1)\lambda + 2\big)^2(2 - \lambda) - (k_1 + 1)\big((n - (k_1 + 2))\lambda + 2\big)}
$$

$$
\cdot \frac{\Big((n - (k_2 + 2))\lambda + 2\Big) - \gamma\lambda(2 - \lambda)\Big((n - 1)\lambda + 2\Big)\Big)^2}{\gamma\big((n-1)\lambda + 2\big)^2(2-\lambda) - (k_2 + 1)\big((n-(k_2 + 2))\lambda + 2\big)}.
$$
(18)

For Eq. (18), we have for the numerator that

$$
\big((n - (k_1 + 2))\lambda + 2\big)\big((n - (k_2 + 2))\lambda + 2\big) < \gamma\lambda(2 - \lambda)\big((n - 1)\lambda + 2\big)^2.
$$

Since $\big((n - (k + 2))\lambda + 2\big)$ increases as $k \to 0$, then it is maximized when $k_1 = 1$ and $k_2 = 0$. This implies

$$
4 < \big((n - 3)\lambda + 2\big)\big((n - 2)\lambda + 2\big) < \gamma\lambda(2 - \lambda)\big((n - 1)\lambda + 2\big)^2.
$$

The expression on the right-hand side $\gamma\lambda(2 - \lambda)\big((n - 1)\lambda + 2\big)^2$ depends on the network size n, the substitution degree λ, and on the effectiveness γ. From Lemma 1, the last parameter γ depends on values of n and λ. This means that finding a lower bound $\bar{\lambda}$ for Case 3 in Proposition 1 depends on n, where n associates with λ to determine the value of γ.

The next proposition shows that the profit of firms is highest for the case of complementary and independent goods, when the R&D network is complete. However, it is important to note that by Lemma 2, the overall size of network $n \leq 2 - 1/\lambda$ when $\lambda < 0$. This implies that although it is natural to think in terms of a grand coalition of firms conducting R&D on a collection of complementary goods, the actual size of the grand coalition may be small. When goods are substitutes and the market size is small, for example, three firms, profit increases as activity level k increases, which means the highest return is obtained when firms form a complete network. However, if the sizes of the market and the substitution degree are not small, there is a level of cooperation $\widehat{k_c}$ at which the profit of all firms is maximized. In [9], Propositions 2 and 7 concern the profit of firms for independent and homogeneous goods. Those propositions are generalized in cases (2) and (3) of the following proposition.

Proposition 2 (Profit of Firms) *For Cournot competition with n firms participating in a k-regular R&D network and zero spillover,*

1. *The profit of firms declines when the substitution degree λ increases.*
2. *When outputs are strategic complements, the profit is increasing as the cooperative activity k increases.*
3. *When outputs are strategic substitutes, if the size of the market n and the substitution degree λ are not small, then there exists an optimal level of cooperative activity $0 < \widehat{k_c} < n - 1$ in which the profits of firms are maximized.*[2]

The proof is given in the Appendix 1.

Proposition 3 states that the complete network where all firms cooperate is the stable network regardless of the size of the network and the competition intensity between firms. This means the incentive of firms to invest in R&D always increases when cooperative links increase. This result extends the results of Goyal and Moraga-Gonzalez for independent and homogeneous goods stated in Propositions 2 and 6. Also, this result confirms results of [8] stated in Proposition 3.1 and Theorem 3.1.

Proposition 3 (Stability of R&D Networks) *For Cournot competition with n firms participating in a k-regular R&D network and zero spillover,*

1. *When outputs are strategic complements, the complete R&D network is uniquely stable network.*
2. *When outputs are strategic substitutes, the complete R&D network is stable network.*

The proof is given in the Appendix 1.

The next proposition shows that total welfare decreases as the substitution degree increases. This indicates that social benefit is maximized when firms are in a differentiated product market, whereas the opposite occurs in product markets where goods are substitutes. Concerning the efficiency of R&D networks, the proposition states that the complete network is the unique efficient network when goods are complements or independent. When goods are substitutes, the effect of activity levels depends on the market size and substitution degree. If market size is small, for example, three firms, then the complete network is efficient. In contrast, if the size

---

[2]The size of the lower bound $\bar{\lambda}$ relies on the market size n. If $n = 3$, the profit of firms increases as the cooperative activity level k increases for all λ. For $n \geq 4$, the profit is maximized at $\widehat{k_c}$ where the threshold value of λ such that the result acquires depends on n—see Table 1 in the Appendix 2.

of the market and the substitution degree are not small, for example, when $n = 5$ and $\lambda \geq 0.9$, the total welfare is maximized at the activity level $0 < \overline{k}_c < n - 1$.

Proposition 4 (Total Welfare) *For Cournot competition with n firms participating in a k-regular R&D network and zero spillover,*

1. *The total welfare declines when the substitution degree λ increases.*
2. *When outputs are strategic complements, the total welfare is increasing as the activity level k increases.*
3. *When outputs are strategic substitutes, if the size of the market n and the substitution degree λ are not small, then there exists a cooperative activity level $\overline{k}_c < n - 1$ in which the total welfare is maximized.[3]*

The proof is given in the Appendix 1.

Note that the impact of the activity levels on the total welfare (Cases 2 and 3 in Proposition 4) generalizes the results of [9] for independent and homogeneous products (Propositions 3 and 8, respectively). However, the result differs from [8] for homogeneous goods. They found that complete network is the unique efficient network when goods are homogeneous (Proposition 3.3). In addition, Proposition 4 when taken with Proposition 3 shows that the conflict between stability and efficiency occurs when goods are substitutes and this depends on the size of the market, with the disparity increasing when the market size increases and as goods become closer substitutes.

4 Bertrand Competition

Under Bertrand competition, it is shown that for the product market stage game, the Nash equilibrium price and quantity for good i are given by

$$p_i^* = \frac{(1 - \lambda)(2 + (2n - 3)\lambda) a}{((2n - 3)\lambda + 2)((n - 3)\lambda + 2)}$$
$$+ \frac{(1 + (n - 2)\lambda)(2 + (n - 2)\lambda)c_i + \lambda(1 + (n - 2)\lambda) \sum_{j \neq i} c_j}{((2n - 3)\lambda + 2)((n - 3)\lambda + 2)} \tag{19}$$

and

$$q_i^* = \left(\frac{1 + (n - 2)\lambda}{(1 - \lambda)(1 + (n - 1)\lambda)} \right) \left[\frac{(1 - \lambda)(2 + (2n - 3)\lambda)a}{((2n - 3)\lambda + 2)((n - 3)\lambda + 2)} \right.$$

---

[3]Note that if the market size $n = 3$, the total welfare increases as the activity level k increases for all λ. If $n \geq 4$, the total welfare is maximized at $\overline{k}_c$ where $0 < \overline{k}_c < n - 1$. The threshold value of λ such that the total welfare is maximized at $\overline{k}_c$ changes with n—see Table 1 in the Appendix 2.

$$-\frac{\left(2+3(n-2)\lambda+(n^2-5n+5)\lambda^2\right)c_i-\lambda\left(1+(n-2)\lambda\right)\sum_{j\neq i}c_j}{\left((2n-3)\lambda+2\right)\left((n-3)\lambda+2\right)}\Bigg].$$

(20)

The Nash equilibrium profit for the i'th firm is derived by substituting Eqs. (19) and (20) into Eq. (4)

$$\pi_i^* = \left(\frac{(1-\lambda)\left(1+(n-1)\lambda\right)}{1+(n-2)\lambda}\right)\Bigg[\frac{(1-\lambda)\left(2+(2n-3)\lambda\right)a}{\left((2n-3)\lambda+2\right)\left((n-3)\lambda+2\right)}$$

$$-\frac{\left(2+3(n-2)\lambda+(n^2-5n+5)\lambda^2\right)c_i-\lambda\left(1+(n-2)\lambda\right)\sum_{j\neq i}c_j}{\left((2n-3)\lambda+2\right)\left((n-3)\lambda+2\right)}\Bigg]^2-\gamma x_i^2.$$

(21)

Now, assume the cooperation between n firms in R&D represented by a symmetric network where each firm has k links (activity level) and spillover is set at zero ($\beta=0$). Let x_i denotes the effort of firm i and x_j is the effort of each firm linked to that firm i. Also, let x_m be the R&D effort of the remaining $n-k-1$ firms in the network. The Nash equilibrium for the R&D stage game is expressed by the following formula:

R&D Effort

$$x^* = \frac{T\left((n-2)\lambda+1\right)(a-\bar{c})}{\gamma\left((2n-3)\lambda+2\right)\left((n-1)\lambda+1\right)\left((n-3)\lambda+2\right)^2-(k+1)\left((n-2)\lambda+1\right)T},$$

(22)

where $T = \left((n-1)(n-3)-(n-2)(k+1)\right)\lambda^2+\left(3(n-2)-k\right)\lambda+2$. Substituting the R&D effort Eq. (22) into the other economic variables yields the equilibrium cost, quantity, and profit functions that are expressed in the following equations:

Cost Function

$$c^* = \frac{\bar{c}\gamma\left((2n-3)\lambda+2\right)\left((n-1)\lambda+1\right)\left((n-3)\lambda+2\right)^2-(k+1)\left((n-2)\lambda+1\right)Ta}{\gamma\left((2n-3)\lambda+2\right)\left((n-1)\lambda+1\right)\left((n-3)\lambda+2\right)^2-(k+1)\left((n-2)\lambda+1\right)T}.$$

(23)

Quantity Function

$$q^* = \frac{\gamma\left(1+(n-2)\lambda\right)\left(2+(2n-3)\lambda\right)\left(2+(n-3)\lambda\right)(a-\bar{c})}{\gamma\left((2n-3)\lambda+2\right)\left((n-1)\lambda+1\right)\left((n-3)\lambda+2\right)^2-(k+1)\left((n-2)\lambda+1\right)T}.$$

(24)

Profit Function

$$\pi^* = \frac{\gamma\left(1+(n-2)\lambda\right)\left[(1-\lambda)\left(1+(n-1)\lambda\right)M-T^2\left(1+(n-2)\lambda\right)\right](a-\overline{c})^2}{\left[\gamma\left((2n-3)\lambda+2\right)\left((n-1)\lambda+1\right)\left((n-3)\lambda+2\right)^2 - (k+1)\left((n-2)\lambda+1\right)T\right]^2},$$
(25)

where $M = \gamma\left(2+(2n-3)\lambda\right)^2\left(2+(n-3)\lambda\right)^2$.

Total Welfare Function

$$TW^* = \frac{\gamma n\left(1+(n-2)\lambda\right)\left[\left(3+(n-4)\lambda\right)\left(1+(n-1)\lambda\right)M-2\left(1+(n-2)\lambda\right)T^2\right](a-\overline{c})^2}{2\left[\gamma\left((2n-3)\lambda+2\right)\left((n-1)\lambda+1\right)\left((n-3)\lambda+2\right)^2 - (k+1)\left((n-2)\lambda+1\right)T\right]^2}.$$
(26)

The following proposition puts a restriction on the values of the R&D effectiveness parameter (γ). This is done in order to have non-negative results in all economic variables under Bertrand competition.

Lemma 3 (Effectiveness Parameter) *For Bertrand competition with n firms participating in a k-regular R&D network and zero spillover, the effectiveness (γ) should satisfy*

$$\gamma > \max \left\{ \frac{(k+1)\left((n-2)\lambda+1\right)Ta}{\overline{c}\left((2n-3)\lambda+2\right)\left((n-1)\lambda+1\right)\left((n-3)\lambda+2\right)^2}, \right.$$
$$\left. \left(\frac{1+(n-2)\lambda}{(1-\lambda)\left(1+(n-1)\lambda\right)}\right)\left[\frac{\left(n^2 - (5+k)n+(2k+5)\right)\lambda^2 + \left(3(n-2)-k\right)\lambda+2}{\left((2n-3)\lambda+2\right)\left((n-3)\lambda+2\right)}\right]^2 \right\}.$$
(27)

The proof is given in the Appendix 1.

The demand function (3) in Bertrand competition is not well defined for homogeneous goods and the second order condition for maximizing the profit function is not satisfied for some values of the substitution degree. Thus some restrictions on the differentiation degree need to be made for this type of competition.[4] These restrictions are captured in the following proposition:

Lemma 4 (Substitution Degree) *Under Bertrand competition with n firms, the values of the substitution degree are restricted as follows:*

---

[4]See also [10, 20] and [12]. These papers stated these conditions.

1. *For an oligopoly market ($n > 2$), the demand function (3) is well defined, if the substitution degree $\lambda \neq 1$ and $\lambda \neq \frac{1}{1-n}$.*
2. *The second order condition for maximizing profit ($\frac{\partial^2 \pi_i}{\partial p_i^2}$) is satisfied if $\lambda > \frac{1}{1-n}$.*

The proof of this proposition is straightforward and a sketch is provided. Case (1) is for two firms in a market, where the equilibria cannot be calculated if goods are perfect complements or substitutes. This case is not discussed in this dissertation since the focus is on oligopolistic markets. For oligopolistic markets, case (2) implies that equilibria cannot be identified analytically under Bertrand competition for perfect substitute goods (homogeneous goods) or if goods are complementary and the substitution degree $\lambda = \frac{1}{1-n}$. Case (3) indicates that for complementary goods ($-1 \leq \lambda < 0$), if the number of firms increases, then the substitution degree should increase to 0 (i.e., independent goods); otherwise, the second order condition for maximizing profit is not satisfied. In other words, the substitution degree is decisive in determining the size of the market. This problem does not appear when goods are independent or substitutes. The substitution degree is restricted only when goods are complements ($\lambda < 0$). To investigate case (3), the demand function (3) is substituted into the profit function (4),

$$\pi_i = (p_i - c_i)\left[\frac{(1-\lambda)a - (1 + (n-2)\lambda)p_i + \lambda \sum_{j \neq i} p_j}{(1-\lambda)\big((n-1)\lambda + 1\big)}\right] - \gamma x_i^2 .$$

The second order condition is satisfied if

$$\frac{-2(1 + (n-2)\lambda)}{(1-\lambda)\big((n-1)\lambda + 1\big)} < 0 .$$

This implies that the substitution degree should satisfy the condition $\lambda > \frac{1}{1-n}$.

In general, the observations in Bertrand competition are similar to those reported for Cournot competition. This does not indicate that the results are the same, but that comparative static behavior of the R&D network in Bertrand competition is similar to that under Cournot competition.

Proposition 5 (R&D Effort) *For Bertrand competition with n firms participating in a k-regular R&D network and zero spillover,*

1. *R&D effort declines when the substitution degree λ increases.*
2. *When prices are strategic substitutes (i.e., when goods are complements), the R&D effort increases as the activity level k increases.*
3. *When prices are strategic complements,*

   (a) *For independent goods, the R&D effort increases as the activity level k increases.*

(b) *For substitute goods where substitution degree is not small for small size of market n, R&D effort declines as the level of cooperative activity k grows.*[5]

The previous proposition indicates that when firms in a market compete by setting the price of their product, the R&D effort varies with different structures of the market and cooperative network. The proposition states that the R&D effort of firms decreases as the substitution degree increases. This means that the R&D investment of firms reaches the highest rate when they are in highly differentiated product markets, whereas the investment is expected to reach the lowest rate when firms operate in highly substitute product markets. The second part of Proposition 5 states that when the prices are strategic complements and goods are independent, the expenditure of firms on R&D attains the maximum amount when R&D collaboration takes the form of a complete network. However, when the goods are substitutes, if the substitution degree is not small when n is small (e.g., if $n = 3$, $\lambda \geq 0.4$), then R&D effort declines as the cooperative activity increases. This means that when goods are substitutes, the expenditure of firms on R&D in a complete network is at its lowest level. As in Cournot competition, determining a lower bound $\overline{\lambda}$ such that item 3 acquires is a problem. If $n = 3$ and $\gamma = 2$, the R&D effort decreases as the activity level k increases for $\lambda \geq 0.3$. If $n > 3$, the R&D effort decreases as the activity level increases for $\lambda > 0.2$.

Proposition 6 (Profit of Firms) *For Bertrand competition with n firms participating in a k-regular R&D network and zero spillover,*

1. *The profit of firms declines when the substitution degree λ increases.*
2. *When prices are strategic substitutes, the profit of firms increases as the activity level k increases.*
3. *When prices are strategic complements,*

   (a) *For independent goods, the profit of firms increases as the activity level k increases.*
   (b) *For substitute goods, if the size of the network n and the substitution degree λ are not small, there is an optimal level of activity, $0 < \widehat{k_b} < n - 1$, for which the profit of firms is maximized.*[6]

Proposition 6 shows that firms prefer investing in highly differentiated product markets since they can obtain higher profits. Also, if the prices are strategic substitutes, firms collectively are expected to form a complete R&D network of cooperation regardless of the size of the market to obtain higher profit and this is true if the prices are strategic complements and goods are independent. Figures in

---

[5] As in Cournot competition, determining a lower bound $\overline{\lambda}$ such that item 3 acquires is based on the number of firms—see Example 2 in the Appendix 2.

[6] Note that, the smallness of λ such that the profit function is maximized at $\widehat{k_b}$ depends on the size of the market n. If $n \leq 5$, the profit of firms increases as the cooperative activity level k increases for all λ. If $n = 7$, the profit is maximized at $\widehat{k_b}$ for $\lambda \geq 0.4$ and if $n = 10$, this result acquires for $\lambda \geq 0.3$ (see Table 1 in Appendix 2).

the Appendix 2 show different values of the activity level for which the profit of firms is maximized with different sizes of the market and substitution degree. For substitute goods, maximizing profit function with respect to activity level k depends on the market size. If the market size is small ($n = 5$), the profit is maximized at $k = n - 1$. However, as long as both the market size and substitution degree are not small (e.g., $n = 7$ and $\lambda \geq 0.5$) there exists a level of collaborative activity $0 < \widehat{k_b} < n - 1$ for which all profits are maximized. This activity level $\widehat{k_b}$ will vary with the size of the market and the degree of substitution between firms' products.

Proposition 7 (Stability of R&D Networks) *For Bertrand competition with n firms participating in a k-regular R&D network and zero spillover, then*

1. *When prices are strategic substitutes or strategic complements where goods are independent (i.e., $\lambda \leq 0$), the complete network K_n is uniquely stable network.*
2. *When prices are strategic complements where goods are substitutes ($\lambda > 0$), the complete network K_n is stable network.*
3. *When goods are homogeneous ($\lambda = 1$), the empty network E_n is uniquely stable network.*

The Proposition 7 concerns the incentive of firms to cooperate in R&D. It shows that for firms, the desirable network in differentiated product markets (goods are complements or independent) is obtained when each two firms in the network agree to collaborate in R&D (a complete network K_n), i.e., the complete network is stable. Also, since the profit increases as the cooperative activity level k increases (item 2 in Proposition 7), the complete network K_n is uniquely stable. When goods are substitutes, the complete network is stable. However, proof of the uniqueness seems a problem since there are arbitrary firms in a network and the proof of the stability needs to consider each link between any two firms. The problem appears if an additional (existing) link between any two firms is added (deleted) where then the equilibria should change and consequently, the profit of these firms will also change.

However, when the goods are homogeneous, the empty network (E_n) is uniquely stable network. This is because each firm seeks to reduce its price to the minimum price (cost of production) to take the market demand. Since this is attained by co-operating in R&D, firms attempt to delete their links to the firm with minimum cost and this yields in the end to an empty network. These results for differentiated and homogeneous goods are consistent with the results of [8] (Proposition 3.2 for homogeneous goods and Theorem 3.1 for differentiated goods).

Proposition 8 (Total Welfare) *For Bertrand competition with n firms participating in a k-regular R&D network and zero spillover,*

1. *Total welfare declines when the substitution degree λ increases.*
2. *When prices are strategic substitutes, the total welfare increases as the activity level k increases.*
3. *When prices are strategic complements,*

(a) *For independent goods, the total welfare increases as the activity level k increases.*

(b) *For substitute goods, if the substitution degree (λ) is not small for small size of market n, then there exists activity level, $0 < \overline{k_b} < n - 1$, in which total welfare is maximized.[7]*

Propositions 7 and 8 highlight that the conflict between stability and efficiency of networks does not appear when prices are strategic substitutes or when prices are strategic complements and goods are independent. However, if goods are substitutes and the substitution degree increases beyond a threshold, then the conflict between stability and efficiency occurs. According to Proposition 8, social welfare under Bertrand competition decreases as the substitution degree λ increases. This indicates that the marginal social benefit for cooperation of firms in R&D is increasing when firms are operating in differentiated product markets, or when firms produce independent or complementary goods. However, when the goods are substitutes and the substitution degree is larger than a threshold value, then the complete network is not efficient. There exists an activity level $0 < \overline{k_b} < n - 1$ for which total welfare is maximized. The threshold value of the substitution degree parameter λ depends on the market size n. If $n = 3$ and $\lambda < 0.9$, then the total welfare increases as the activity level k increases. However, if $n = 5$, the total welfare is maximized at $\overline{k_b}$ for $\lambda \geq 0.5$. If $n = 10$, the result acquires for $\lambda \geq 0.2$. Table 1 in Appendix 2 shows values of $\overline{k_b}$ for different sizes of n and λ. Goyal and Joshi [8] use a different model from the model presented in this paper. However, they found that

Table 1 Activity levels for which the profit and total welfare are maximized in Cournot and Bertrand competition

Type of competition	Size of the market	Effectiveness γ	Substitution degree							
			$\lambda = 0.1$		$\lambda = 0.3$		$\lambda = 0.5$		$\lambda = 0.9$	
Activity level			$\widehat{k}$	$\overline{k}$	$\widehat{k}$	$\overline{k}$	$\widehat{k}$	$\overline{k}$	$\widehat{k}$	$\overline{k}$
Cournot	3	1	2	2	2	2	2	2	2	2
	5	2	4	4	4	4	4	4	2	2
	7	2	6	6	6	6	4	4	4	4
	10	2	9	9	7	7	6	6	6	5
	20	2	19	19	13	12	12	11	13	10
Bertrand	3	2	2	2	2	2	2	2	2	0
	5	2	4	4	4	4	4	2	4	2
	7	2	6	6	6	6	4	4	6	2
	10	3	9	9	7	7	6	5	8	4
	20	3	18	18	12	12	12	10	16	9

---

[7]The threshold value of the substitution degree λ such that item 3 holds depends on the market size n. If $n = 3$ and $\lambda < 0.9$, then the total welfare increases as the activity level k increases. If $n = 5$, the total welfare is maximized at $\overline{k_b}$ for $\lambda \geq 0.5$. If $n = 10$, the result acquires for $\lambda \geq 0.2$. Table 1 in Appendix 2 shows values of $\overline{k_b}$ for different sizes of n and λ.

the complete and the empty networks are not efficient see [8], Proposition 3.4. Our results complement and complete their results.

5 Comparison of Cournot and Bertrand Equilibria

When comparing Cournot and Bertrand competition for differentiated goods, it is found that R&D investments of firms for complementary goods are higher when there is price competition. However, if goods are substitutes, firms spend more on R&D under Cournot competition. The result in R&D effort is consistent with [19] for a duopoly market where goods are substitutes. Also, Bertrand competition always yields lower prices and higher output. This indicates that the consumer surplus is always higher in Bertrand competition irrespective of the size of the market and the structure of the market and the network. This result coincides with results of [4, 19, 20] for duopoly, and with the result of [12] for oligopoly. Moreover, if goods are complements, it is found that Bertrand competition is preferable for firms because profit is higher than under Cournot. However, if goods are substitutes, Cournot competition is more profitable. This result is consistent with [20] for a duopoly market and with [10] for an oligopoly market. We also find that when goods are complements, the total welfare under Bertrand competition is higher than under Cournot competition. When goods are substitutes and the substitution degree is high (e.g., $n = 10$ and $\lambda = 0.9$), the social benefit is high under Cournot competition, but only if the structure of cooperation in the R&D networks is dense (i.e., firms have a large number of collaborators k relative to the size of the network). Figure 1 illustrates the comparison between Cournot and Bertrand competitions.

Proposition 9 (Comparison Between Cournot and Bertrand Competition) *For n firms participating in a k-regular R&D network with zero spillover, the following are the differences between Cournot and Bertrand equilibria for complementary and substitute goods:*

1. R&D effort of firms is higher and the cost is lower in Bertrand competition if goods are complements, whereas if goods are substitutes, the R&D effort is higher and the cost is lower in Cournot competition.

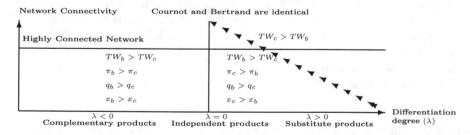

Fig. 1 Cournot versus Bertrand for differentiated goods

2. *For all kinds of products, the output under Bertrand competition is higher than under Cournot competition. This indicates that the consumer surplus is higher and the price is lower in Bertrand competition than in Cournot competition.*
3. *Profit of firms is higher in Bertrand competition if goods are complements, whereas if goods are substitutes, the profit in Cournot is higher.*
4. *Total welfare is higher in Bertrand competition if goods are complements and this statement is true for substitute goods except if the substitution degree is very high and the activity level k is not small.*[8]

The proof is given in the Appendix 1.

6   Conclusion

This paper examines the impact of product differentiation on an R&D network. This paper shows that there are differences between Cournot and Bertrand competition in terms R&D effort, profit, and total welfare. The variations between the two competition models depend on values of the substitution degree and number of cooperative links. If products are complements, Bertrand competition is a dominant strategy, because profit of firms under Bertrand competition always exceeds profit under Cournot competition. Firms are expected to invest more highly in R&D than under Cournot competition. However, if goods are substitutes, firms' investment and return are high under Cournot competition. For production of firms, price is lower and production is higher under Bertrand competition than under Cournot regardless of number of firms and types of products (complement or substitute), and structure of the network.

This result confirms that consumer surplus is always higher under Bertrand competition, which is consistent with [12]. In terms of social benefits, total welfare is higher under Bertrand competition except if goods are substitutes and the network is dense, in the sense that the majority of firms are connected. This result stands apart from the conventional wisdom that Bertrand competition delivers higher welfare where goods are homogeneous [20]. This result extends prior works by Singh and Vives [20], Hackner [10], and Hsu and Wang [12] and indicates that the structure of the network is a matter in the comparison between Cournot and Bertrand in terms of total welfare. Our results can be seen as supporting [11], which have provided a comprehensive study on the impact of product differentiation on the non-tournament R&D model and complementing the study in [8] for R&D networks. Our paper contrasts with these results, showing that when goods are substitutes, there is a threshold effect that determines the relative welfare optimality of Bertrand versus Cournot competition.

---

[8]The smallness of the activity level depends on the sizes of the market and the substitution degree. Example 3 in the Appendix 2 shows that for ten firms when $\lambda = 0.9$, the total welfare in Bertrand is higher than in Cournot if the activity level $k < 4$.

Acknowledgements This research was supported by King Saud University, Deanship of Scientific Research, College of Science Research Center. We are grateful for useful suggestions and comments made by Shravan Luckraz and an anonymous referee.

Appendix 1

Proof of Lemma 1

Proof This restriction on values of the effectiveness is made by satisfying the following conditions:

1. The effort function should be non-negative ($x^* \geq 0$). Then, from the effort function (12), the effectiveness (γ) should satisfy

$$\gamma > \frac{(k+1)\big((n-(k+2))\lambda + 2\big)}{\big((n-1)\lambda + 2\big)^2 (2-\lambda)} .$$

2. The cost function (13) should give non-negative results. This is obtained if

$$\gamma \geq \frac{(k+1)\big((n-(k+2))\lambda + 2\big)a}{\bar{c}\big((n-1)\lambda + 2\big)^2 (2-\lambda)} .$$

3. The second order condition for maximizing profit function ($\frac{\partial^2 \pi}{\partial x^2} < 0$) is satisfied if

$$\gamma > \left[\frac{(n-(k+2))\lambda + 2}{\big((n-1)\lambda + 2\big)(2-\lambda)} \right]^2 .$$

Since $a > \bar{c}$, the requirement set by condition (1) is achieved by satisfied condition (2). When goods are either complements or independent (i.e., $\lambda \leq 0$), then condition (2) is necessary and sufficient for R&D effort to be positive valued and for the second order condition to be satisfied. However, when goods are substitutes, conditions (2) and (3) are together important in determining the level that γ must attain for investment in R&D to be positive and satisfy the second order conditions set by the Nash equilibrium of the R&D stage game.   ∎

Proof of Lemma 2

Proof For appropriate values of R&D effectiveness γ determined by Lemma 1, the denominator of output function (14) is always positive. However, the numerator,

$\gamma(2 - \lambda)\big((n - 1)\lambda + 2\big)(a - \overline{c})$, can be either zero or negative. To show this, set $\lambda = 2/(1 - n)$ and substitute this into the quantity function (14). The result is zero, which means firms do not produce in the market. When $\lambda < 2/(1 - n)$, the quantity function gives negative outcomes. To see this, assume $\lambda = 3/(1-n) < 2/(1-n)$ and without loss of generality set $a - \overline{c} = 1$ ($a - \overline{c}$ must always be positive; otherwise, firms will not produce). By substituting $\lambda = 3/(1-n)$ into the output function (14),

$$q^* = \frac{-\gamma(2n + 1)}{\gamma(2n + 1) + (k + 1)(n - 3k - 4)} < 0.$$

Hence, $\lambda > \frac{2}{1-n}$. ∎

Proof of Proposition 1

Proof

1. Let $x^*_{\lambda_1}$ and $x^*_{\lambda_2}$ be the R&D investments of firms in two distinct R&D networks, $G_k(\lambda_1)$ and $G_k(\lambda_2)$ that are, respectively, associated with substitution degrees λ_1 and λ_2, such that $\lambda_1 < \lambda_2$ and satisfy Lemma 2. Without loss of generality, we set λ_1 and $\lambda_1 > 0$ and $(a - \overline{c}) = 1$. Assume that γ satisfies Lemma 1, so that R&D is non-negative and maximizes profit at the Nash equilibrium. To compare $x^*_{\lambda_1}$ and $x^*_{\lambda_2}$, we take the difference between them,

$$x^*_{\lambda_1} - x^*_{\lambda_2} = \frac{\gamma}{\gamma\big((n-1)\lambda_1 + 2\big)^2(2 - \lambda_1) - (k+1)\big((n - (k+2))\,\lambda_1 + 2\big)}$$
$$\cdot \left[\frac{(2 - \lambda_2)\big((n - (k+2))\lambda_1 + 2\big)\big((n - 1)\lambda_2 + 2\big)^2}{\Big(\gamma\big((n-1)\lambda_2 + 2\big)^2(2 - \lambda_2) - (k+1)\big((n - (k+2))\,\lambda_2 + 2\big)\Big)} \right.$$
$$\left. - \frac{(2 - \lambda_1)\big((n - (k+2))\lambda_2 + 2\big)\big((n - 1)\lambda_1 + 2\big)^2}{\Big(\gamma\big((n - 1)\lambda_2 + 2\big)^2(2 - \lambda_2) - (k+1)\big((n - (k+2))\,\lambda_2 + 2\big)\Big)} \right].$$

We require for $\lambda_i > 2/(1 - n)$, $i = 1, 2$, each part of $x^*_1 - x^*_2$ to be positive. Since $\lambda_1 < \lambda_2$,

$$(2 - \lambda_2)\big((n - (k + 2))\lambda_1 + 2\big) \leq (2 - \lambda_1)\big((n - (k + 2))\lambda_2 + 2\big), \tag{28}$$

where the right-hand side of the inequality (28) equals the left-hand side if the activity level $k = n-1$. This means as $k \to 0$, the difference between the left- and right-hand sides of (28) increases. Also, $((n - 1)\lambda_2 + 2)^2 > ((n - 1)\lambda_1 + 2)^2$ and the difference between the right- and left-hand sides depends on the size of the market n and the difference between λ_1 and λ_2. For each n and $\lambda_1 < \lambda_2$, it can be found that

$$(2 - \lambda_2)\big((n - (k + 2))\lambda_1 + 2\big)\big((n - 1)\lambda_2 + 2\big)^2$$
$$> (2 - \lambda_1)\big((n - (k + 2))\lambda_2 + 2\big)\big((n - 1)\lambda_1 + 2\big)^2.$$

To prove that this inequality holds, consider the following three cases in terms of the values of the cooperative activity level k:

(a) If $k = n - 1$, the right- and left-hand sides of (4.14) are equal. Since $\lambda_1 < \lambda_2$, then $((n - 1)\lambda_2 + 2)^2 > ((n - 1)\lambda_1 + 2)^2$. This means the previous inequality is proved.

(b) If $k = n - 2$, $\big(n - (k + 2)\lambda_1 + 2\big) = \big(n - (k + 2)\lambda_2 + 2\big)$. Since $0 < \lambda_1 < \lambda_2 \leq 1$, and $1 \leq (2 - \lambda_i) < 2$ for $i = 1, 2$ and $((n - 1)\lambda_2 + 2)^2 > ((n - 1)\lambda_1 + 2)^2$. This proves the inequality.

(c) If $k < n - 2$, since $\lambda_1 < \lambda_2$, then $\big(n - (k + 2)\lambda_1 + 2\big) < \big(n - (k + 2)\lambda_2 + 2\big)$.

Now, since the difference between the two sides in (28) increases as $k \to 0$, let $k = 0$. The task is completed by showing that

$$(2 - \lambda_2)\big((n - 2)\lambda_1 + 2\big)\big((n - 1)\lambda_2 + 2\big)^2 > (2 - \lambda_1)\big((n - 2)\lambda_2 + 2\big)\big((n - 1)\lambda_1 + 2\big)^2.$$

If the difference between λ_1 and λ_2 is large where $0 < \lambda_1 < \lambda_2$, then $((n - 1)\lambda_2 + 2)^2 > ((n - 1)\lambda_1 + 2)^2$ and this proves the last inequality. If the difference between λ_1 and λ_2 is small, then the difference between $(2 - \lambda_2)$ and $(2 - \lambda_1)$ and between $((n - 2)\lambda_2 + 2)$ and $((n - 2)\lambda_1 + 2)$ is small. This means that $(2 - \lambda_2)\big((n - 2)\lambda_1 + 2\big) < (2 - \lambda_1)\big((n - 2)\lambda_2 + 2\big)$ with a small difference that depends on the network size n. Since $((n - 1)\lambda_2 + 2)^2 > ((n - 1)\lambda_1 + 2)^2$, the inequality holds.[9]

2. The second result is proved by showing that when goods are complements or independent, then $x^*(G_{k_1}) > x^*(G_{k_2})$ where G_{k_1} and G_{k_2} are symmetric networks with k_1 and k_2 links, respectively. Assume $k_1 > k_2$ and substitute k_1 and k_2 into the effort function (12). Without loss of generality, assume $(a - \bar{c}) = 1$. To compare between $x^*(G_{k_1})$ and $x^*(G_{k_2})$,

$$x^*(G_{k_1}) - x^*(G_{k_2}) = \frac{k_1 - k_2}{\gamma\big((n - 1)\lambda + 2\big)^2(2 - \lambda) - (k_1 + 1)\big((n - (k_1 + 2))\lambda + 2\big)}$$
$$\cdot \left[\frac{\big((n - (k_1 + 2))\lambda + 2\big)\big((n - (k_2 + 2))\lambda + 2\big)}{\gamma\big((n - 1)\lambda + 2\big)^2(2 - \lambda) - (k_2 + 1)\big((n - (k_2 + 2))\lambda + 2\big)} \right.$$
$$\left. - \frac{\gamma\lambda(2 - \lambda)\big((n - 1)\lambda + 2\big)^2}{\gamma\big((n - 1)\lambda + 2\big)^2(2 - \lambda) - (k_2 + 1)\big((n - (k_2 + 2))\lambda + 2\big)} \right].$$

---

[9]This result does not occur if the substitution degree $\lambda \to 1$ and the cooperative activity level k is small. However, the main aim of item 1 in Proposition 1 is to show that in each network structure, the R&D effort in a weakly competitive market ($\lambda \leq 0$) is higher than in a competitive market ($\lambda > 0$).

In the case when goods are independent ($\lambda = 0$), the fraction (18) becomes

$$x^*(G_{k_1}) - x^*(G_{k_2}) = \frac{k_1 - k_2}{\left[4\gamma - (k_1 + 1)\right]\left[4\gamma - (k_2 + 1)\right]} > 0,$$

as $k_1 > k_2$ and the effectiveness $\gamma > \frac{n}{4}$ (from Lemma 2). This implies $x^*(G_{k_1}) > x^*(G_{k_2})$.

For complementary goods ($\frac{2}{1-n} < \lambda < 0$), the numerator in the fraction (18) is positive where $\left((n - (k_i + 2))\lambda + 2\right) > 0$ for $i = 1, 2$ (via Lemma 2). The denominator is always positive from Lemma 1. This means $x^*(G_{k_1}) > x^*(G_{k_2})$ and then the result (2) follows.

3. When goods are substitutes where $\lambda > 0$ is not small for n small, the proof is completed by showing that

$$\left((n - (k_1 + 2))\lambda + 2\right)\left((n - (k_2 + 2))\lambda + 2\right) < \gamma\lambda(2 - \lambda)\left((n - 1)\lambda + 2\right)^2.$$

Since $\lambda > 0$, the term on the left-hand side in the previous inequality decreases as the cooperative links k_1 and k_2 increase. Also, for any $k_2 < k_1 \leq n - 1$,

$$\left((n - (k_1 + 2))\lambda + 2\right)\left((n - (k_2 + 2))\lambda + 2\right) < \left((n - 1)\lambda + 2\right)^2.$$

On the right-hand side, if $0 < \lambda \leq 1$, then $1 \leq (2 - \lambda) < 2$. Also, the right-hand side is affected by smallness of λ, particularly when n is small. Thus, for $\lambda > 0$ not small where the effectiveness γ is chosen to be sufficiently large (Lemma 1), the inequality is proved and this implies that $x^*(G_{k_1}) < x^*(G_{k_2})$. ∎

Proof of Proposition 2

Proof

1. Let π_1^* and π_2^* be profits of firms that are associated with differentiation degrees λ_1 and λ_2, respectively, such that $\lambda_1 < \lambda_2$. Let $\lambda_1, \lambda_2 > 0$ and $(a - \bar{c}) = 1$. It is found that

$$sign\{\pi_1^* - \pi_2^*\} = sign\left\{\gamma^2\left[\gamma^2(2-\lambda_1)^2(2-\lambda_2)^2\left((n-1)\lambda_1 + 2\right)^2\left((n-1)\lambda_2 + 2\right)^2\right.\right.$$

$$\cdot\left((n - 1)(\lambda_2 - \lambda_1)\left((n - 1)(\lambda_1 + \lambda_2) + 4\right)\right)$$

$$+ (2 - \lambda_1)\left((n - 1)\lambda_1 + 2\right)^2\left((n - (k + 2))\lambda_2 + 2\right)$$

$$\cdot \Big[((n - (k+2))\lambda_2 + 2) P_1 - (2 - \lambda_1)(k+1) P_2 \Big]$$

$$+ (2 - \lambda_2)\big((n-1)\lambda_2 + 2\big)^2 \big((n - (k+2))\lambda_1 + 2\big)$$

$$\Big[(2 - \lambda_2)(k+1) P_3 - \big((n - (k+2))\lambda_1 + 2\big) P_4 \Big] \Big] \Big\},$$

$$\tag{29}$$

where

$$P_1 = \gamma(2 - \lambda_1)\big((n-1)\lambda_1 + 2\big)^2 - 2(k+1)\big((n - (k+2))\lambda_1 + 2\big),$$

$$P_2 = 2\gamma(2 - \lambda_2)\big((n-1)\lambda_2 + 2\big)^2 - (k+1)\big((n - (k+2))\lambda_2 + 2\big),$$

$$P_3 = 2\gamma(2 - \lambda_1)\big((n-1)\lambda_1 + 2\big)^2 - (k+1)\big((n - (k+2))\lambda_1 + 2\big)$$

and

$$P_4 = \gamma(2 - \lambda_2)\big((n-1)\lambda_2 + 2\big)^2 - 2(k+1)\big((n - (k+2))\lambda_2 + 2\big).$$

Since $\lambda_1 < \lambda_2$, then $\big((n-1)\lambda_2+2\big)^2 > \big((n-1)\lambda_1+2\big)^2$. Also, from Lemma 2, the expressions P_1, P_2, P_3, and P_4 are positive.

Note that since the negative term in P_1 is multiplied by 2 where $\lambda_1 < \lambda_2$, then $P_2 > P_1$. Similarly, we have $P_3 > P_4$ in most cases. Also, these expressions P_i for $i = 1, \ldots, 4$ are multiplied by other terms and that makes the square brackets that contain them small or have different *signs*. In contrast, the first expression

$$\gamma^2(2 - \lambda_1)^2(2 - \lambda_2)^2\big((n-1)\lambda_1 + 2\big)^2$$

$$\big((n-1)\lambda_2 + 2\big)^2\Big((n-1)(\lambda_2 - \lambda_1)\big((n-1)(\lambda_1 + \lambda_2) + 4\big)\Big)$$

is a multiplication of square and positive functions. This implies $\pi_1^* > \pi_2^*$ which means the profit function decreases as the substitution degree λ increases.

2. The second result can be verified by showing that when goods are complements or independent, if the activity levels $k_1 > k_2$, then $\pi^*(G_{k_1}) > \pi^*(G_{k_2})$. Let $(a - \bar{c}) = 1$ and substitute k_1 and k_2 into the profit function (15).

If goods are independent ($\lambda = 0$), then

$$\pi^*(G_{k_1}) - \pi^*(G_{k_2}) = \frac{\gamma(4\gamma - 1)\big[\big(4\gamma - (k_2 + 1)\big)^2 - \big(4\gamma - (k_1 + 1)\big)^2\big]}{\big(4\gamma - (k_1 + 1)\big)^2\big(4\gamma - (k_1 + 1)\big)^2}.$$

From Lemma 1, the effectiveness $\gamma > \frac{n}{4}$. Since $k_2 < k_1 \leq n - 1$, $\pi^*(G_{k_2}) < \pi^*(G_{k_1})$.

If goods are complements ($\frac{2}{1-n} < \lambda < 0$), the required task is to show that $\pi^*(G_{k_1}) > \pi^*(G_{k_2})$ for $k_1 > k_2$. It is found that

$$sign\big(\pi^*(G_{k_1}) - \pi^*(G_{k_2})\big) = sign\Big\{\big((n-1)\lambda + 2\big)$$

$$\Big[\gamma(2-\lambda)\big((n-1)\lambda + 2\big)\big(\lambda((n+k_2)\lambda - 2(2k_2+1))$$

$$+ (4-\lambda)\big((n-(k_1+2))\lambda + 2\big)\big)$$

$$- 2\big((n-(k_1+2))\lambda + 2\big)\big((n-(k_2+2))\lambda + 2\big)\Big]$$

$$+ (2-\lambda)\big(\lambda(3+k_1+k_2-n) - 2\big)\big)$$

$$\cdot \big((k_1+1)\big((n-(k_1+2))\lambda + 2\big)$$

$$+ (k_2+1)\big((n-(k_2+2))\lambda + 2\big)\Big\}. \qquad (30)$$

For $\lambda < 0$, then $\lambda\big(\lambda(n+k_2)-2(2k_2+1)\big) > 0$ and $(4-\lambda)\big((n-(k_1+2))\lambda+2\big) > 0$. Also, since $\frac{2}{1-n} < \lambda < 0$, then $2 < (2-\lambda) < \frac{2n}{n-1}$. Since $\big((n-1)\lambda+2\big) > 0$ and the effectiveness γ is large, particularly when $\lambda < 0$ (Lemma 1), then

$$\gamma(2-\lambda)\big((n-1)\lambda+2\big)\big(\lambda((n+k_2)\lambda - 2(2k_2+1))$$

$$+ (4-\lambda)\big((n-(k_1+2))\lambda + 2\big)\big)$$

$$- 2\big((n-(k_1+2))\lambda + 2\big)\big((n-(k_2+2))\lambda + 2\big) > 0.$$

The expression $(2-\lambda)\big(\lambda(3+k_1+k_2-n)-2\big)\big((k_1+1)\big((n-(k_1+2))\lambda+2\big)+(k_2+1)\big((n-(k_2+2))\lambda+2\big)\big) < 0$ since $\lambda(3+k_1+k_2-n)-2 < 0$. However, because of the term $\gamma\big((n-1)\lambda+2\big)^2$ with large effectiveness γ, $\pi^*(G_{k_1}) > \pi^*(G_{k_2})$.

3. When goods are substitutes ($\lambda > 0$), the proof is completed by showing that $\pi^*(G_{n-1}) < \pi^*(G_{n-2})$. Substituting $k = n-1$ and $k = n-2$ into the profit function (15) results in

$$\pi^*(G_{n-1}) = \frac{\gamma(a-\overline{c})^2\big[\gamma\big((n-1)\lambda+2\big)^2 - 1\big]}{\big[\gamma\big((n-1)\lambda+2\big)^2 - n\big]^2},$$

$$\pi^*(G_{n-2}) = \frac{\gamma(a-\overline{c})^2\big[\gamma(2-\lambda)^2\big((n-1)\lambda+2\big)^2 - 4\big]}{\big[\gamma(2-\lambda)\big((n-1)\lambda+2\big)^2 - 2(n-1)\big]^2}.$$

For comparison between the previous two profits, it is found that

$$sign\left(\pi^*(G_{n-1}) - \pi^*(G_{n-2})\right) = sign\Big\{\gamma\left((n-1)\lambda + 2\right)^2$$
$$\cdot [\gamma\left((n-1)\lambda + 2\right)^2\left((2n-1)\lambda^2 - 4n\lambda + 8\right)$$
$$- \left(n^2\lambda^2 - 4\lambda(n^2 - n + 1) + 4(2n+1)\right)]$$
$$+ 4(2n-1)\Big\}. \tag{31}$$

When $\lambda > 0$, the sign of the expression $\left((2n-1)\lambda^2 - 4n\lambda + 8\right)$ changes with the size of the market n and the substitution degree λ. If n and λ are not small, it is found that $(2n-1)\lambda^2 - 4n\lambda + 8 < 0$ and because of the term $\gamma\left((n-1)\lambda + 2\right)^2$, $\pi^*(G_{n-1}) - \pi^*(G_{n-2}) < 0$.

∎

Proof of Proposition 3

Proof

1. When output is a strategic complement ($\lambda \leq 0$), Proposition 2 (item 2) shows that the profit function (15) increases as the activity level k increases. This implies that the complete network K_n is uniquely stable.
2. When output is a strategic substitute ($0 < \lambda \leq 1$), the required task is to prove that $\pi^*(G_{n-1}) - \pi^*(G_{n-1} - ij) > 0$.

   Here the procedure to prove the stability of the complete network is adopted from Goyal and Moraga-Gonzalez. The second condition of the stability of networks (Definition 1) is satisfied since no link can be added to the complete network. Thus, the proof is completed by showing the first condition. This means in the complete network, a single link between any two firms will be deleted and if the gain of at least one of the two firms after deletion decreases, the complete network is stable. Let G_{n-1} denote to the complete network, the profit of each firm i in that network is

$$\pi^*(G_{n-1}) = \frac{\gamma(a - \bar{c})^2\left[\gamma\left((n-1)\lambda + 2\right)^2 - 1\right]}{\left[\gamma\left((n-1)\lambda + 2\right)^2 - n\right]^2}.$$

Assume that the link between firms i and j is removed. Then, the resulting network is $G_{n-1} - ij$. Now, in the network $G_{n-1} - ij$, there are two types of firms. The first type is firms i and j linked to $n-2$ firms and the second type is the remaining firms with each having $n-1$ links. For symmetric solutions, the effort of the remaining firms is denoted by x_p and the effort of each firm in that group is denoted by x_r. The cost function for firm i in $G_{n-1} - ij$ is $c_i = \bar{c} - x_i - (n-2)x_p$

and by symmetry, the cost c_j can be found. Also, the cost function for firm r is
$c_r = \bar{c} - x_r - x_i - x_j - (n-3)x_p$.

Thus the profit function for firm i is

$$\pi_i^*(G_{n-1}-ij) = \left[\frac{(2-\lambda)(a-\bar{c}) + 2x_i - \lambda(n-1)x_j + (n-2)(2-\lambda)x_p}{(2-\lambda)\big((n-1)\lambda+2\big)} \right]^2 - \gamma x_i^2 .$$

Similarly, the profit of firm j can be found. Also, the profit of each firm r from the group of firms that have $n-1$ links is

$$\pi_r^*(G_{n-1}-ij) = \left[\frac{(2-\lambda)(a-\bar{c}) + 2x_i + 2x_j + (2-\lambda)x_r + (n-3)(2-\lambda)x_p}{(2-\lambda)\big((n-1)\lambda+2\big)} \right]^2 - \gamma x_r^2 .$$

From the profit function for firm i in the network $G_{n-1} - ij$, the first order condition is

$$2\big((2-\lambda)(a-\bar{c}) + 2x_i - \lambda(n-1)x_j + (n-2)(2-\lambda)x_p\big)$$
$$- \gamma\big((2-\lambda)\big((n-1)\lambda+2\big)\big)^2 x_i = 0 .$$

Similarly, the first order condition for firm r is

$$\big((2-\lambda)(a-\bar{c}) + 2x_i + 2x_j + (2-\lambda)x_r$$
$$+ (n-3)(2-\lambda)x_p\big) - \gamma(2-\lambda)\big((n-1)\lambda+2\big)^2 x_r = 0 .$$

By symmetry $x_i = x_j$ and $x_r = x_p$, and by doing some substitution, the R&D efforts of the different types of firms are

$$x_i^*(G_{n-1} - ij) = \frac{a-\bar{c}}{Q} 2\gamma(2-\lambda)A^2\big[\gamma(2-\lambda)^2 A - 2\big] ,$$

$$x_p^*(G_{n-1} - ij) = \frac{a-\bar{c}}{Q}\big[\gamma^2(2-\lambda)^4 A^3 - 4(n+1)\big] ,$$

where

$$Q = \gamma^3(2-\lambda)^4 A^5 - \gamma^2(2-\lambda)^2 A^3\big(\lambda(\lambda-4)(n-2) + 4n\big)$$
$$- 4\gamma A\big(\lambda(n-1)^2 - 4\big) + 4(n^2 - n - 2)$$

and $A = (n-1)\lambda + 2$. By substituting these R&D effort equations into the profit function of firm i in the network $G_{n-1} - ij$, the optimal profit function of firm i becomes

$$\pi_i^*(G_{n-1} - ij) = \frac{(a - \bar{c})^2}{((n-1)\lambda + 2)^2 B_1^2}$$

$$\cdot \left[\left(B_1 + 2\gamma(2 - (n-1)\lambda)A^2(\gamma(2-\lambda)^2 A - 2) + B_2 \right)^2 \right.$$

$$\left. - 4\gamma^2(2-\lambda)^2 A^6(\gamma(2-\lambda)^2 A - 2)^2 \right],$$

where

$$B_1 = \gamma^3(2-\lambda)^4 A^5 - \gamma^2(2-\lambda)^2 A^3(\lambda(\lambda - 4)(n-2) + 4n)$$

$$- 4\gamma A(\lambda(n-1)^2 - 4) + 4(n^2 - n - 2),$$

and

$$B_2 = (n-2)(\gamma^2(2-\lambda)^4 A^3 - 4(n+1)).$$

Without loss of generality, let $(a - \bar{c}) = 1$. To compare between the profits of firm i in networks G_{n-1} and $G_{n-1} - ij$,

$$\pi_i^*(G_{n-1}) - \pi_i^*(G_{n-1} - ij)$$

$$= \frac{1}{\left((\gamma A^2 - n) A B_1 \right)^2}$$

$$\left[\gamma A^2 \left(B_1^2(\gamma A^2 - 1) + 4\gamma^2(2-\lambda)^2 A^4(\gamma A^2 - n)^2(\gamma(2-\lambda)^2 A - 2)^2 \right) \right.$$

$$\left. - (\gamma A^2 - n)^2 (B_1 + 2\gamma A^2(2 - (n-1)\lambda)(\gamma(2-\lambda)^2 A - 2) + B_2)^2 \right].$$

$$(32)$$

For homogeneous goods ($\lambda = 1$) with $\gamma = 1$, the previous fraction becomes

$$\pi_i^*(G_{n-1}) - \pi_i^*(G_{n-1} - ij) = \frac{4n^7 + 15n^6 - 4n^5 - 20n^4 + 30n^3 + 8n^2 + 2n + 3}{(n^2 + n + 1)^2(n^3 + 4n^2 - 2n + 1)^2}.$$

$$(33)$$

The last fraction is positive which implies that $\pi_i^*(G_{n-1}) > \pi_i^*(G_{n-1} - ij)$. Note that, for homogeneous goods, the result has been shown by Goyal and Moraga-Gonzalez [9].

Now, the required task is to prove the statement for $0 < \lambda < 1$. Note that, the term $2 - (n-1)\lambda < 0$ if $\lambda > \frac{2}{n-1}$. This means that as the network size n increases, the term $2\gamma A^2(2 - (n-1)\lambda)(\gamma(2-\lambda)^2 A - 2)$ is negative for most values of λ. Also, the expression B_1 decreases because of the negative terms where $\lambda(\lambda - 4)(n - 2) + 4n > 0$ and $\lambda(n - 1)^2 - 4 < 0$ if $\lambda < \frac{4}{(n-1)^2}$ which means that as n increases the previous term is positive for most values of λ.

In contrast, the first expression on the second line of Eq. (32), $\gamma A^2\left(B_1^2(\gamma A^2 - 1) + 4\gamma(2-\lambda)^2 A^4(\gamma A^2 - n)^2(\gamma(2-\lambda)^2 A - 2)^2\right)$, is a sum and multiplication of square functions. This means that the numerator of $\pi_i^*(G_{n-1}) - \pi_i^*(G_{n-1} - ij)$ is positive which implies $\pi_i^*(G_{n-1}) > \pi_i^*(G_{n-1} - ij)$.

■

Proof of Proposition 4

Proof

1. The total welfare function is defined by the quantity and the profit functions

$$TW^* = \underbrace{\frac{(1-\lambda)}{2}\sum_{i=1}^{n}q_i^{*\,2} + \frac{\lambda}{2}\left(\sum_{i=1}^{n}q_i^*\right)^2}_{Consumer\ surplus(CS)} + \underbrace{\sum_{i=1}^{n}\pi_i^*}_{Industry\ profit}.$$

Since the profit function declines with growing the substitution degree, the proof is completed by showing that the quantity function also decreases with growing the substitution degree.

Assume that there are two quantities of output q_1^* and q_2^*, for two different values of the substitution degree λ_1 and λ_2, respectively, such that $0 < \lambda_1 < \lambda_2$. The required task is to prove that $q_1^* > q_2^*$. Without loss of generality, assume $(a - \bar{c}) = 1$ and for comparison between the two quantities, it is found that $q_1^* > q_2^*$ if and only if

$$\gamma(n-1)(\lambda_2 - \lambda_1)(2 - \lambda_1)(2 - \lambda_2)\big((n-1)\lambda_2 + 2\big)\big((n-1)\lambda_1 + 2\big) >$$

$$(k+1)\Big[(2-\lambda_1)\big((n-1)\lambda_1 + 2\big)\big((n-(k+2))\lambda_2 + 2\big)$$

$$- (2-\lambda_2)\big((n-1)\lambda_2 + 2\big)\big((n-(k+2))\lambda_1 + 2\big)\Big].$$

$$(34)$$

Since $\lambda_1 < \lambda_2$, then $(\lambda_2 - \lambda_1) > 0$. This means that the left-hand side of the inequality (34) is positive. Also, for each n, the term on the right hand of that inequality is maximized when the activity level $k = 0$. However, the right-hand side consists of subtraction of two positive functions compared to the expression on the left-hand side, which is multiplication of positive functions. This proves the inequality (34) which means that the quantity of production declines as the substitution degree increases.

2. The second result is proved by showing that when the output is a strategic complement ($\lambda \leq 0$), if the activity levels $k_1 > k_2$, then $TW^*(G_{k_1}) > TW^*(G_{k_2})$. Let $(a - \bar{c}) = 1$ and substitute k_1 and k_2 into the total welfare function (16).

If goods are independent ($\lambda = 0$), then

$$TW(G_{k_1}) - TW(G_{k_2}) = \frac{n\gamma(6\gamma - 1)\left[\left(4\gamma - (k_2 + 1)\right)^2 - \left(4\gamma - (k_1 + 1)\right)^2\right]}{\left[4\gamma - (k_1 + 1)\right]^2 \left[4\gamma - (k_2 + 1)\right]^2}.$$

Since $k_1 > k_2$ where $\gamma > \frac{n}{4}$ (Lemma 1), then $TW^*(G_{k_1}) > TW^*(G_{k_2})$.

Now, this result is shown for complementary goods ($\frac{2}{1-n} < \lambda < 0$) for any $k_1 > k_2$. It is found that

$$sign\{TW(G_{k_1}) - TW(G_{k_2})\}$$

$$= sign\Bigg\{\left((n-1)\lambda + 2\right)\Bigg[2\gamma(2 - \lambda)\left((n-1)\lambda + 2\right)$$

$$\cdot \left[\lambda\Big((k_2 + 1)(\lambda - 2)\left((n-1)\lambda + 3\right) + \left((n - (k_2 + 2))\lambda + 2\right)\Big)\right.$$

$$+ \left((n - (k_1 + 2))\lambda + 2\right)\Big((2 - \lambda)\left((n-1)\lambda + 3\right) + \lambda\Big)\Bigg]$$

$$- 4\left((n - (k_1 + 2))\lambda + 2\right)\left((n - (k_2 + 2))\lambda + 2\right)\Bigg]$$

$$+ (2 - \lambda)\left(\lambda(3 + k_1 + k_2 - n) - 2\right)\left((n-1)\lambda + 3\right)$$

$$\cdot \Big((k_1 + 1)\left((n - (k_1 + 2))\lambda + 2\right)$$

$$+ (k_2 + 1)\left((n - (k_2 + 2))\lambda + 2\right)\Big)\Bigg\}. \tag{35}$$

Since $\lambda < 0$, then $(k_2 + 1)(\lambda - 2)\left((n-1)\lambda + 3\right) < 0$. For any $0 \leq k_2 < n - 1$, $\left((n-1)\lambda + 3\right) > \left((n - (k_2 + 2))\lambda + 2\right)$ and this implies $(k_2 + 1)(\lambda - 2)\left((n-1)\lambda + 3\right) + \left((n - (k_2 + 2))\lambda + 2\right) < 0$. This means $\lambda\Big((k_2 + 1)(\lambda - 2)\left((n-1)\lambda + 3\right) + \left((n - (k_2 + 2))\lambda + 2\right)\Big) > 0$. Also, it can be found that $\left((n - (k_1 + 2))\lambda + 2\right)\Big((2 - \lambda)\left((n-1)\lambda + 3\right) + \lambda\Big) > 0$.

Other terms in (35) are negative, but since $\left((n-1)\lambda + 2\right)^2 > \left((n - (k_1 + 2))\lambda + 2\right)\left((n - (k_2 + 2))\lambda + 2\right)$ and the effectiveness γ is large, particularly when $\lambda < 0$ (Lemma 1), then $TW(G_{k_1}) > TW(G_{k_2})$.

3. When the output is a strategic substitute ($\lambda > 0$), the proof is completed by showing that $TW^*(G_{n-1}) < TW^*(G_{n-2})$. Let $(a - \bar{c}) = 1$, then substituting $k = n - 1$ and $k = n - 2$ into the total welfare function (16), respectively, yields

$$TW^*(G_{n-1}) = \frac{n\gamma \left[\gamma\,(2 + \lambda(n-1))^2\,(3 + (n-1)\lambda) - 2\right]}{2\left[\gamma\,(2 + (n-1)\lambda)^2 - n\right]^2},$$

$$TW^*(G_{n-2}) = \frac{n\gamma \left[\gamma(2-\lambda)^2\,(2 + \lambda(n-1))^2\,(3 + (n-1)\lambda) - 8\right]}{2\left[\gamma(2-\lambda)\,(2 + (n-1)\lambda)^2 - 2(n-1)\right]^2}.$$

For comparison between the two total welfare functions,

$$TW^*(G_{n-1}) - TW^*(G_{n-2}) = \frac{n\gamma}{2\left[\gamma\,(2 + \lambda(n-1))^2 - n\right]^2}$$

$$\cdot \frac{\left[\gamma\,(2 + (n-1)\lambda)^2\,(F_1 + F_2) - 8(2n + 1)\right]}{\left[\gamma(2-\lambda)\,(2 + \lambda(n-1))^2 - 2(n-1)\right]^2},$$

where

$$F_1 = 2\gamma\,(2 + (n-1)\lambda)^2 \left(n(n-1)\lambda^3\right.$$

$$\left. + (1 + 3n - 2n^2)\lambda^2 - 2(n+3)\lambda + 12\right) - 8(2 + (n-1)\lambda)$$

and

$$F_2 = ((n-1)\lambda + 3)\left(4(1 - 2n) + 4n^2\lambda - n^2\lambda^2\right).$$

Assume $\lambda > 0$ and n is not small, it is found that $F_1 < 0$ because $n(n-1)\lambda^3 + (1 + 3n - 2n^2)\lambda^2 - 2(n+3)\lambda + 12 < 0$, whereas $F_2 > 0$. Since the effectiveness γ is sufficiently large (Lemma 1) with the term $(2 + \lambda(n-1))^2$, $F_1 - F_2 < 0$. This implies $TW^*(G_{n-1}) - TW^*(G_{n-2}) < 0$ and the result follows.

∎

Proof of Proposition 3

Proof The restriction of the effectiveness is based on satisfying the following conditions:

1. The R&D effort (22) should be non-negative. Thus, it can be found that

$$\gamma > \frac{(k+1)\big((n-2)\lambda+1\big)T}{\big((2n-3)\lambda+2\big)\big((n-1)\lambda+1\big)\big((n-3)\lambda+2\big)^2}.$$

2. The cost function (23) should be non-negative which means effectiveness

$$\gamma \geq \frac{(k+1)\big((n-2)\lambda+1\big)Ta}{\overline{c}\big((2n-3)\lambda+2\big)\big((n-1)\lambda+1\big)\big((n-3)\lambda+2\big)^2}.$$

3. The second order condition for maximizing profit function ($\frac{\partial^2 \pi}{\partial x^2} < 0$) is satisfied
   if

$$\gamma > \frac{1+(n-2)\lambda}{(1-\lambda)(1+(n-1)\lambda)}$$

$$\left[\cdot \frac{\big(n^2-(5+k)n+(2k+5)\big)\lambda^2 + \big(3(n-2)-k\big)\lambda+2}{\big((2n-3)\lambda+2\big)\big((n-3)\lambda+2\big)} \right]^2.$$

Since $a > \overline{c}$, condition (2) dominates condition (1). ∎

Proof of Proposition 9

Proof The proof of this proposition depends on the functions of all variables in both competitions for symmetric networks. Also, these statements are proved for the case when the network is complete ($k = n - 1$). The comparison between Cournot and Bertrand competition has studied by Singh and Vives [20], Hackner [10], Hsu and Wang [12], and Qiu [19]. The difference between former works and that in this paper is that the comparison of Cournot and Bertrand competition is made using R&D networks. Moreover, items 1, 2, and 3 in Proposition 9 are not affected by the activity level k, but they are affected by the values of the differentiation degree λ. Item 4 in the proposition is affected by both the differentiation degree λ and the activity level k. The total welfare in Cournot exceeds the total welfare in Bertrand if the substitution degree is very high and the activity level is not small and this is not consistent with [12]. Therefore, it is sufficient to prove item 4 by assuming that $k = n - 1$.

For simplicity, the statements are also shown by assuming that $(a - \overline{c}) = 1$. Note that to have positive equilibrium output in Cournot competition, the substitution degree should satisfy $\lambda > \frac{2}{1-n}$ (Lemma 2). Also, in Bertrand competition $\lambda > \frac{1}{1-n}$ because of the second order condition for maximizing the profit function (Lemma 4). This means that $\lambda > \frac{1}{1-n}$ is the sufficient condition to ensure appropriate equilibria in both competitions. Moreover, to have non-negative results for all economic variables, assume that effectiveness (γ) is appropriate for both competitions together.[10]

1. The task here is to prove that for complementary goods, the R&D effort in Bertrand competition is higher than in Cournot competition. The opposite relationship holds for substitute goods. Denote the Cournot and Bertrand equilibrium efforts by x_c^* and x_b^*, respectively. By substituting $k = n - 1$ in the R&D effort functions in Cournot and Bertrand (Eqs. (12) and (22), respectively), this yields

$$x_c^* = \frac{1}{\gamma\big((n-1)\lambda+2\big)^2 - n},$$

$$x_b^* = \frac{(1-\lambda)\big((n-2)\lambda+1\big)}{\gamma\big((n-1)\lambda+1\big)\big((n-3)\lambda+2\big)^2 - n(1-\lambda)\big((n-2)\lambda+1\big)}.$$

To compare between the two efforts, calculate the difference between them which yields

$$x_c^* - x_b^* = \frac{\gamma\lambda^3(n-1)^2}{\gamma\big((n-1)\lambda+2\big)^2 - n}$$

$$\cdot \frac{(n-2)\lambda+2}{\gamma\big((n-1)\lambda+1\big)\big((n-3)\lambda+2\big)^2 - n(1-\lambda)\big((n-2)\lambda+1\big)}. \tag{36}$$

For complementary goods ($\frac{1}{1-n} < \lambda < 0$), the previous fraction (36) is negative because $\lambda^3 < 0$. This means $x_c^* < x_b^*$ which indicates that the R&D effort is higher and the cost of production is lower in Bertrand competition than in Cournot competition.[11] For substitute goods ($\lambda > 0$), the fraction (36) is positive i.e., $x_c^* > x_b^*$ which means that the R&D effort is lower and the cost is higher in Bertrand than in Cournot if goods are substitutes.

2. Let q_c^* and q_b^* denote quantities of production under Cournot and Bertrand competition, respectively. For any value of the substitution degree such that $\lambda > \frac{1}{1-n}$, proved that $q_b^* > q_c^*$. Let $k = n - 1$ in q_c^* and q_b^*. To compare the quantities of production under Cournot and Bertrand competition, calculate $q_c^* - q_b^*$. This yields

$$q_c^* - q_b^* = -\frac{\gamma\lambda^2(n-1)}{\gamma\big((n-1)\lambda+2\big)^2 - n}$$

$$\cdot \frac{\gamma\big((n-3)\lambda+2\big)\big((n-1)\lambda+2\big) - n\big((n-2)\lambda+1\big)}{\gamma\big((n-1)\lambda+1\big)\big((n-3)\lambda+2\big)^2 - n\big((n-2)\lambda+1\big)(1-\lambda)}. \tag{37}$$

---

[10]From Lemmas 1 and 3, let $\gamma > max\ \{\gamma_c, \gamma_b\}$ where γ_c and γ_b are the effectiveness in Cournot and Bertrand competition, respectively.

[11]Since $c^* = \bar{c} - (k+1)x^*$, if $x_b^* > x_c^*$, then $c_b^* < c_c^*$ where c_c^* and c_b^* are the equilibrium cost in Cournot and Bertrand competition, respectively.

For $\lambda > \frac{1}{1-n}$, each term in the previous fraction is positive and because the effectiveness γ is large,

$$\gamma\big((n-3)\lambda+2\big)\big((n-1)\lambda+2\big) > n\big((n-2)\lambda+1\big).$$

This makes the previous fraction always negative, which indicates that $q_c^* < q_b^*$. This means that the output and then consumer surplus are higher, and the price is lower in Bertrand competition than in Cournot competition.[12]

3. Let π_c and π_b denote the profit functions in Cournot and Bertrand, respectively. By substituting $k = n - 1$ in the profit functions in both competitions, calculate $\pi_c - \pi_b$ which yields

$$
\begin{aligned}
sign\{\pi_c-\pi_b\} = sign\Big\{&\Big(\big[\gamma^2\big((n-1)\lambda+1\big)\big((n-1)\lambda+2\big)^2\big((n-2)\lambda+2\big)\big((n-3)\lambda+2\big)^2 \\
&+ n(n-2)(1-\lambda)\big((n-2)\lambda-2\big)\big((n-2)\lambda+1\big)\big]\lambda^3(n-1)^2 \\
&- \gamma\Big[\big((n-1)\lambda+1\big)^2\big((n-3)\lambda+2\big)^4 \\
&- (1-\lambda)^2\big((n-1)\lambda+2\big)^4\big((n-2)\lambda+1\big)^2\Big]\Big)\gamma^2\Big\}.
\end{aligned}
$$

The required proof is that $\pi_c < \pi_b$ for $\lambda < 0$. Thus, the proof is completed by showing the following inequality

$$
\begin{aligned}
\lambda^3(n-1)^2\Big[&\gamma^2\big((n-1)\lambda+1\big)\big((n-1)\lambda+2\big)^2\big((n-2)\lambda+2\big)\big((n-3)\lambda+2\big)^2 \\
&+ n(n-2)\big((n-2)\lambda-2\big)(1-\lambda)\big((n-2)\lambda+1\big)\Big] < \\
\gamma\Big[&\big((n-1)\lambda+1\big)^2\big((n-3)\lambda+2\big)^4 - (1-\lambda)^2\big((n-1)\lambda+2\big)^4\big((n-2)\lambda+1\big)^2\Big].
\end{aligned}
$$

$$(38)$$

Note that the term $n(n-2)\big((n-2)\lambda-2\big)(1-\lambda)\big((n-2)\lambda+1\big) < 0$ if $\lambda < \frac{2}{n-2}$, which depends on n. This means that if $\lambda < 0$, the previous term is always negative. However, for any $\frac{1}{1-n} < \lambda < 1$,

$$
\begin{aligned}
&\gamma^2\big((n-1)\lambda+1\big)\big((n-1)\lambda+2\big)^2\big((n-2)\lambda+2\big)\big((n-3)\lambda+2\big)^2 \\
&+ n(n-2)\big((n-2)\lambda-2\big)(1-\lambda)\big((n-2)\lambda+1\big) > 0,
\end{aligned}
$$

because of large effectiveness γ and quadrature. With the term $\lambda^3(n-1)^2 < 0$, the left-hand side of the inequality (38) is negative and lower than the right-

---

[12]Consumer surplus $CS^* = \frac{1-\lambda}{2}\sum_{i=1}^{n} q_i^{*\,2} + \frac{\lambda}{2}(\sum_{i=1}^{n} q_i^*)^2$ and since $q_c^* < q_b^*$, then $CS_c^* < CS_b^*$ where CS_c^* and CS_b^* are consumer surplus in Cournot and Bertrand competition, respectively.

hand side. This yields $\pi_c^* < \pi_b^*$. If goods are substitutes ($\lambda > 0$), for increasing n, $n(n-2)\big((n-2)\lambda - 2\big)(1-\lambda)\big((n-2)\lambda + 1\big)$ is positive for a large range of λ. Also, the left-hand side of the inequality (38) is positive and higher than when $\lambda < 0$. Whereas the right-hand side is reduced by the term $(1-\lambda)^2\big((n-1)\lambda+2\big)^4\big((n-2)\lambda+1\big)^2$ which makes the right-hand side lower than the left-hand side. This yields $\pi_c^* > \pi_b^*$ which means that the profit in Cournot competition is higher than under Bertrand competition if goods are substitutes.

4. For total welfare, let TW_c^* and TW_b^* be the total welfare in Cournot and Bertrand, respectively. Let $k = n - 1$, then

$$TW_c^* = \frac{n\gamma\left[\gamma\big(2+(n-1)\lambda\big)^2\big((n-1)\lambda+3\big)-2\right]}{2\left[\gamma\big(2+(n-1)\lambda\big)^2-n\right]^2},$$

$$TW_b^* = \frac{n\gamma\big(1+(n-2)\lambda\big)\left[\begin{array}{c}\gamma\big(1+(n-1)\lambda\big)\big(2+(n-3)\lambda\big)^2\big(3+(n-4)\lambda\big)\\-2\big(1+(n-2)\lambda\big)(1-\lambda)^2\end{array}\right]}{2\left[\gamma\big(1+(n-1)\lambda\big)\big(2+(n-3)\lambda\big)^2-n\big(1+(n-2)\lambda\big)(1-\lambda)\right]^2}.$$

Assuming that goods are complements ($\lambda < 0$), prove that $TW_c^* < TW_b^*$. To show this, the following inequality should be substantiated:

$$\left[\gamma\big(2+\lambda(n-1)\big)^2\big((n-1)\lambda+3\big)-2\right]\left[\gamma\big(1+(n-1)\lambda\big)\big(2+(n-3)\lambda\big)^2-n(1-\lambda)\big(1+(n-2)\lambda\big)\right]^2$$

$$< \big(1+(n-2)\lambda\big)\left[\gamma\big(1+(n-1)\lambda\big)\big(2+(n-3)\lambda\big)^2\big(3+(n-4)\lambda\big)-2\big(1+(n-2)\lambda\big)(1-\lambda)^2\right]$$

$$\cdot\left[\gamma\big(2+\lambda(n-1)\big)^2-n\right]^2. \tag{39}$$

Note that each term in square bracket in the previous inequality (39) is positive because of quadrature or because of $\lambda > \frac{1}{1-n}$ and the large R&D effectiveness parameter (γ). Also, each term is an increasing function with respect to the substitution degree (λ). This means for small $\lambda < 0$, each term in the previous inequality is small. Assume n large, then the term

$$\left[\gamma\big(1+(n-1)\lambda\big)\big(2+(n-3)\lambda\big)^2 - n(1-\lambda)\big(1+(n-2)\lambda\big)\right]^2, \tag{40}$$

on the left-hand side of the inequality (39) is very small because of $n(1-\lambda)$ compared to $[\gamma\big(2+\lambda(n-1)\big)^2 - n]^2$ on the right-hand side. This makes the left-hand side small, then $TW_c^* < TW_b^*$.

For substitute goods ($\lambda > 0$), assume that goods are high substitutes. Then each term in the previous inequality is high and the previous term (40) becomes very large. This reverses the above inequality, then $TW_b^* < TW_c^*$. ∎

Appendix 2

Example 1 (Activity Levels) For $a = 120$ and $\bar{c} = 100$, the following example provides cooperative activity levels $\widehat{k}$ and $\bar{k}$ for which the profit and total welfare are, respectively, maximized in Cournot and Bertrand competition for different values of substitution degree. Note that the effectiveness (γ) is for $\lambda \in [0.1, 0.9]$.

If $n > 3$, then the profit function and total welfare for substitute goods are maximized at different intermediate activity levels $\widehat{k}$ and $\bar{k}$, respectively. In Cournot, see Propositions 2 and 4 on pages 185 and 186, respectively. In Bertrand, see Propositions 6 and 8 on pages 190 and 191, respectively.

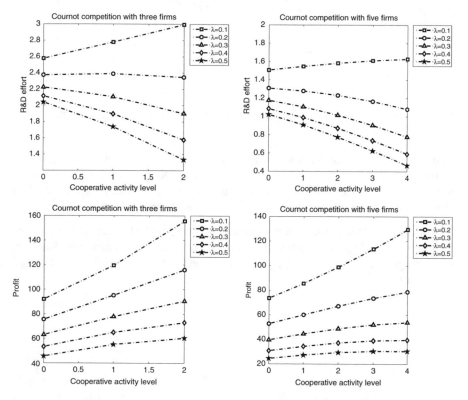

Fig. 2 R&D effort and profit of firms under Cournot competition with three and five firms

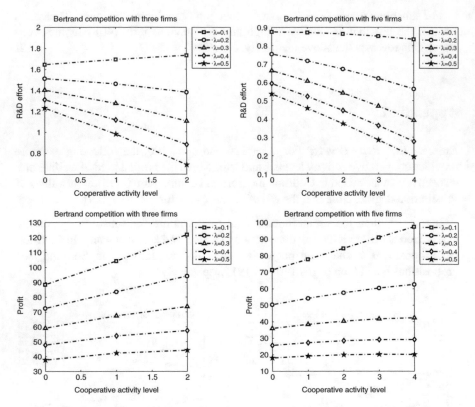

Fig. 3 R&D effort and profit of firms under Bertrand competition with three and five firms

Example 2 (R&D Effort for Small Size of Market) In Cournot and Bertrand competition, let $a = 120$ and $\bar{c} = 100$. Figures 2 and 3 show R&D effort for $\lambda \geq 0.1$. With $n = 3$, $\gamma = 2$ in Cournot competition and $\gamma = 3$ in Bertrand competition. With $n = 5$, $\gamma = 3$ in Cournot competition and $\gamma = 5$ in Bertrand competition.

R&D effort does not decrease when the activity level increases if the sizes of the market n and the substitution degree λ are small as in this example. See item (3) in Proposition 1 under Cournot competition and in Proposition 5 under Bertrand competition.

Table 2 Comparison between Cournot and Bertrand competition

Variables	Effort (x)	Cost (c)	Quantity (q)	Price (p)	Profit (π)	Total welfare (TW)
$\lambda < 0$	$x_c < x_b$	$c_c > c_b$	$q_c < q_b$	$p_c > p_b$	$\pi_c < \pi_b$	$TW_c < TW_b$
$\lambda = 0$	$x_c = x_b$	$c_c = c_b$	$q_c = q_b$	$p_c = p_b$	$\pi_c = \pi_b$	$TW_c = TW_b$
$\lambda > 0$	$x_c > x_b$	$c_c < c_b$	$q_c < q_b$	$p_c > p_b$	$\pi_c > \pi_b$	$TW_c > TW_b$ if $\lambda \to 1$ and the network is dense

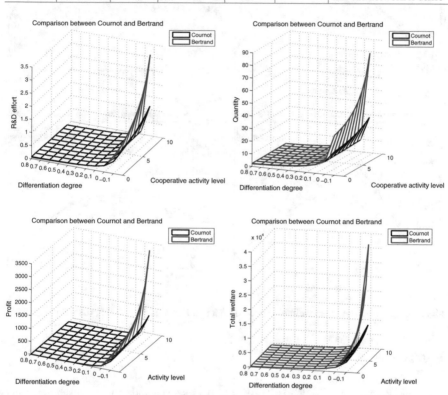

Fig. 4 Comparison between profit under Cournot and Bertrand for $\lambda = 0.8, 0.85, 0.9$, and 0.95

Example 3 (Cournot and Bertrand Competition) Table 2 summarizes the impact of λ. Figure 4 shows the outcomes under Cournot and Bertrand competition for $a = 120, \bar{c} = 100, n = 10, -0.1 \leq \lambda \leq 0.8$, and $\gamma = 21$. Figure 5 shows the total welfare under Cournot and Bertrand competition for $a = 120, \bar{c} = 100, n = 10$, $0.8 \leq \lambda \leq 0.95$, and $\gamma = 2$.

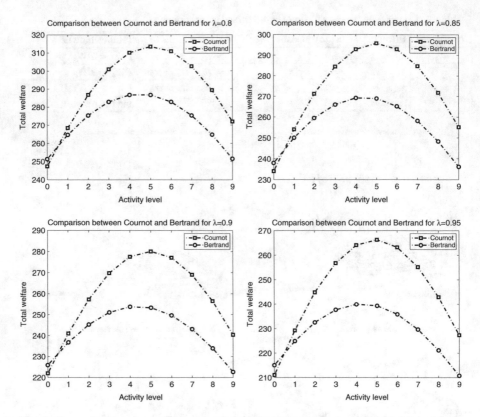

Fig. 5 Comparison between welfare under Cournot and Bertrand for λ = 0.8, 0.85, 0.9, and 0.95

References

1. Amir, R., and Jin, J. Y., 2001. Cournot and Bertrand equilibria compared substitutability, complementarity and concavity. *International Journal of Industrial Organization* 19, 303–317.
2. D'Aspremont, C., and Jacquemin, A., 1988. Cooperative and Noncooperative R&D in Duopoly with Spillovers. American Economic Review 78, 1133–1137.
3. Bulow, J. I., Geanakoplos, J. D., and Klemperer, P. D., 1985. Multimarket Oligopoly: Strategic Substitutes and Complements. Journal of Political Economy 93, 488–511.
4. Cheng, L., 1985. Comparing Bertrand and Cournot Equilibria: A Geometric Approach. RAND Journal of Economics 16, 146–152.
5. Choi, J. P., 1993. Cooperative R&D with Product Market Competition. International Journal of Industrial Organization 11, 553–571.
6. De Bondt, R., Slaets, P., and Cassiman, B., 1992. The degree of spillovers and the number of rivals for maximum effective R&D. International Journal of Industrial Organization 10, 35–54.
7. Deroian, F., 2008. Dissemination of spillovers in cost-reducing alliances. Research in Economics 62, 34–44.
8. Goyal, S., and Joshi, S., 2003. Networks of Collaboration in Oligopoly. Games and Economic Behavior 43 (1), 57–85.

9. Goyal, S., and Moraga-Gonzalez, J. L., 2001. R&D Networks. Rand Journal of Economics 32, 686–707.
10. Hackner, J., 2000. A Note on Price and Quantity Competition in Differentiated Oligopolies. Journal of Economic Theory 93, 233–239.
11. Hinloopen, J. and Vandekerckhove, J. (2011) Product Market Competition and Investments in Cooperative R&D. BE Journal of Economic Analysis and Policy 11(1) Art. 55
12. Hsu, J., and Wang, X. H., 2005. On Welfare under Cournot and Bertrand Competition in Differentiated Oligopolies. Review of Industrial Organization 27, 185–191.
13. Jackson, M., 2008. Social and Economic Networks. Princeton, NJ: Princeton University Press.
14. Jackson, M. O., Wolinsky, A., 1996. A strategic model of social and economic networks. Journal of Economic Theory 71(1), 44–74.
15. Kamien, M., Muller, E., and Zang, I., 1992. Research Joint Ventures and R&D Cartels. American Economic Review 82, 1293–1306.
16. Katz, M., 1986. An Analysis of Cooperative Research and Development. Rand Journal of Economics 17, 527–543.
17. Konig, M. D., Battiston, S., Napoletano, M., and Schweitzer, F., 2012. The efficiency and stability of R&D networks. Games and Economic Behavior 75, 694–713.
18. Leahy, D., and Neary, P., 1997. Public Policy Towards R&D in Oligopolistic Industries. American Economic Review 87, 642–662.
19. Qiu, L. D., 1997. On the Dynamic Efficiency of Bertrand and Cournot Equilibria. Journal of Economic Theory 75(1), 213–229.
20. Singh, N., and Vives, X., 1984. Price and Quantity Competition in a Differentiated Duopoly. Rand Journal of Economics 15, 546–554.
21. Spence, M., 1984. Cost Reduction, Competition, and Industry Performance. Econometrica 52, 101–121.
22. Song, H., and Vannetelbosch, V., 2007. International R&D Collaboration Networks. The Manchester School 75(6), 742–766.
23. Suzumura, K., 1992. Cooperative and Noncooperative R&D in an Oligopoly with Spillovers. American Economic Review 82, 1307–1320.
24. Yi, S.-S., 1996. The Welfare Effects of Cooperative R&D in Oligopoly with Spillovers. Review of Industrial Organization 11, 681–698.
25. Westbrock, B., 2010. Natural concentration in industrial research collaboration. RAND Journal of Economics 41 (2), 351–371.
26. Zu, L., Dong, B., Zhao, X., and Zhang, J., 2011. International R&D Networks. Review of International Economics 19, 325–340.

Part IV
Numerical Methods in Games and Dynamic Games

A Global Optimization Approach to Nonzero Sum Six-Person Game

Rentsen Enkhbat, S. Batbileg, N. Tungalag, A. Anikin, and A. Gornov

Abstract The nonzero sum six-person game has been examined. It is well known that nonzero sum n-person game reduces to a nonconvex optimization problem (Enkhbat et al, IGU Ser Mat 20:109121, 2017). Based on Mills' result (Mills, J Soc Ind Appl Math 8(2):397–402, 1960), we derive a sufficient condition for a Nash equilibrium. To find a Nash equilibrium numerically, we apply the curvilinear multistart algorithm (Gornov and Zarodnyuk, Mach Learn Data Anal 10(1):1345–1353, 2014) developed for nonconvex optimization. The algorithm was tested numerically on six-person game. The game data was generated by "Gamut" (website: http://gamut.stanford.edu/db/generators.html). The number of variables of the reduced optimization problems was varied from 29 to 33. In all cases, Nash equilibriums were found.

Keywords Nash equilibrium · Nonzero sum game · Mixed strategies · Curvilinear multistart algorithm

1 Introduction

Game theory as a part of operations research plays an important role in science and technology as well as in decision theory. There are a lot works devoted to game theory [7, 9–14]. Most of them deal with two person games or nonzero sum two person games. The two-person nonzero sum game was studied in [4, 5, 12] based on D.C programming [1]. The three-person game was examined in [3] by global optimization techniques. So far, less attention has been paid to computational aspects of game theory, especially n-person game.

R. Enkhbat (✉) · S. Batbileg · N. Tungalag
National University of Mongolia, Ulaanbaatar, Mongolia

A. Anikin · A. Gornov
Matrosov Institute for System Dynamics and Control Theory, SB of RAS, Irkutsk, Russian Federation

© Springer Nature Switzerland AG 2020
D. Yeung et al. (eds.), *Frontiers in Games and Dynamic Games*, Annals of the
International Society of Dynamic Games 16,
https://doi.org/10.1007/978-3-030-39789-0_7

219

This paper examines nonzero sum six-person game. The paper is organized as follows. In Sect. 2, we formulate nonzero sum n- person game and show that it can be formulated as a global optimization problem with nonconvex constraints. An algorithm to finding a Nash equilibrium for nonzero sum six-person game was proposed in Sect. 3. Section 4 is devoted to computational results.

2   Nonzero Sum n-Person Game

Consider the n-person game in mixed strategies with matrices $(A_q, \quad q = 1, 2, \ldots, n)$ for players $1, 2, \ldots, n$.

$$A_q = \left(a^q_{i_1 i_2 \ldots i_n} \right), \quad q = 1, 2, \ldots, n$$

$$i_1 = 1, 2, \ldots, k_1, \ldots, i_n = 1, 2, \ldots, k_n.$$

Denote by D_p the set

$$D_p = \{ u \in R^q \mid \sum_{i=1}^p u_i = 1, \ u_i \geq 0, \ i = 1, \ldots, p \}, \ p = k_1, k_2, \ldots, k_n.$$

A mixed strategy for player 1 is a vector $x^1 = (x^1_1, x^1_2, \ldots, x^1_{k_1}) \in D_{k_1}$, where x^1_i represents the probability that player 1 uses a strategy i. Similarly, the mixed strategies for q-th player are $x^q = (x^q_1, x^q_2, \ldots, x^q_{k_q}) \in D_{k_q}$, $q = 1, 2, \ldots, n$. Their expected payoffs are given by for 1-th person :

$$f_1(x^1, x^2, \ldots, x^n) = \sum_{i_1=1}^{k_1} \sum_{i_2=1}^{k_2} \cdots \sum_{i_n=1}^{k_n} a^1_{i_1 i_2 \ldots i_n} x^1_{i_1} x^2_{i_2} \ldots x^n_{i_n}$$

and for q-th person

$$f_q(x^1, x^2, \ldots, x^n) = \sum_{i_1=1}^{k_1} \sum_{i_2=1}^{k_2} \cdots \sum_{i_n=1}^{k_n} a^q_{i_1 i_2 \ldots i_n} x^1_{i_1} x^2_{i_2} \ldots x^n_{i_n},$$

$$q = 1, 2, \ldots, n.$$

Definition 1 A vector of mixed strategies $\tilde{x}^q \in D_{k_q}$, $q = 1, 2, \ldots, n$ is a Nash equilibrium if

$$\begin{cases} f_1(\tilde{x}^1, \tilde{x}^2, \dots, \tilde{x}^n) \geq f_1(x^1, \tilde{x}^2, \dots, \tilde{x}^n), \ \forall x^1 \in D_{k_1} \\ \cdots \quad \cdots \quad \cdots \quad \cdots \quad \cdots \\ f_q(\tilde{x}^1, \tilde{x}^2, \dots, \tilde{x}^n) \geq f_q(\tilde{x}^1, \dots, \tilde{x}^{q-1}, x^q, \tilde{x}^{q+1}, \dots, \tilde{x}^n), \ \forall x^q \in D_{k_q} \\ \cdots \quad \cdots \quad \cdots \quad \cdots \quad \cdots \quad \cdots \\ f_n(\tilde{x}^1, \tilde{x}^2, \dots, \tilde{x}^n) \geq f_n(\tilde{x}^1, \tilde{x}^2, \dots, x^n), \ \forall x^n \in D_{k_n}. \end{cases}$$

By definition, we have

$$f_1(\tilde{x}^1, \tilde{x}^2, \dots, \tilde{x}^n) = \max_{x^1 \in D_{k_1}} f_1(x^1, \tilde{x}^2, \dots \quad \dots, \tilde{x}^n),$$

$$\cdots \quad \cdots \quad \cdots \quad \cdots$$

$$f_q(\tilde{x}^1, \tilde{x}^2, \dots, \tilde{x}^n) = \max_{x^q \in D_{k_q}} f_q(\tilde{x}^1, \tilde{x}^2, \dots, \tilde{x}^{q-1}, x^q, \tilde{x}^{q+1}, \dots, \tilde{x}^n),$$

$$\cdots \quad \cdots \quad \cdots \quad \cdots$$

$$f_n(\tilde{x}^1, \tilde{x}^2, \dots, \tilde{x}^n) = \max_{x^n \in D_{k_n}} f_n(\tilde{x}^1, \tilde{x}^2, \dots, \tilde{x}^{n-1}, x^n).$$

Denote by

$$\sum_{i_1=1}^{k_1} \sum_{i_2=1}^{k_2} \cdots \sum_{i_{q-1}=1}^{k_{q-1}} \sum_{i_{q+1}=1}^{k_{q+1}} \cdots \sum_{i_n=1}^{k_n} a^q_{i_1 i_2 \dots i_n} x^1_{i_1} x^2_{i_2} \dots x^{q-1}_{i_{q-1}} x^{q+1}_{i_{q+1}} \dots x^n_{i_n} \triangleq$$

$$\triangleq \varphi_{i_q}(x^1, x^2, \dots, x^{q-1}, x^{q+1}, \dots, x^n) = \varphi_{i_q}(x|x^q)$$

$$i_q = 1, 2, \dots, k_q, \quad q = 1, 2, \dots, n.$$

For further purpose, it is useful to formulate the following statement.

Theorem 1 ([2]) *A vector strategy* $(\tilde{x}^1, \tilde{x}^2, \dots, \tilde{x}^n)$ *is a Nash equilibrium if and only if*

$$f_q(\tilde{x}) = \varphi_{i_q}(\tilde{x}|\tilde{x}^q) \tag{1}$$

for

$$\tilde{x} = (\tilde{x}^1, \tilde{x}^2, \dots, \tilde{x}^n)$$

$$i_q = 1, 2, \dots, k_q,$$

$$q = 1, 2, \dots, n.$$

The proof is similarly [8].

Theorem 2 ([2]) *A mixed strategy* $\tilde{x}$ *is a Nash equilibrium for the nonzero sum n-person game if and only if there exists vector* $\tilde{p} \in R^n$ *such that vector* $(\tilde{x}, \tilde{p})$ *is a solution to the following bilinear programming problem :*

$$\max_{(x,p)} F(x, p) = \sum_{q=1}^{n} f_q(x^1, x^2, \dots, x^n) - \sum_{q=1}^{n} p_q \tag{2}$$

subject to :

$$\varphi_{i_q}(x|x^q) \le p_q, \quad i_q = 1, 2, \dots, k_q. \tag{3}$$

The proof is the same as in [2].

3   The Curvilinear Multistart Algorithm

In order to solve problem (2)–(3), we use curvilinear multistart algorithm in [6]. The algorithm was originally developed for solving box-constrained optimization problems; therefore, we convert our problem from the constrained to unconstrained form using penalty function techniques. For each equality constraint $g(x) = 0$, we construct a simple penalty function $\hat{g}(x) = g^2(x)$. For each inequality constraint $q(x) \le 0$, we also construct the corresponding penalty function as follows:

$$\hat{q}(x) = \begin{cases} 0, & \text{if } q(x) \le 0, \\ q^2(x), & \text{if } q(x) > 0. \end{cases}$$

Thus, we have the following box-constrained optimization problem:

$$\hat{f}(x) = f(x) + \frac{\gamma}{2} \sum_{i} \hat{g}_i(x) + \frac{\gamma}{2} \sum_{j} \hat{q}_j(x) \to \min_{X},$$

$$X = \left\{ x \in \mathbb{R}^n \mid \underline{x_i} \le x_i \le \overline{x_i}, \ i = 1, \dots, n \right\},$$

where γ is a penalty parameter, $\underline{x}$ and $\overline{x}$ are the lower and upper bounds. For original x-variables the constraint is the box $[0, 1]$; for p-variables box constraints are $[0, \overline{p_q}]$. Values of $\overline{p_q}$ are chosen from some intervals. An initial value of a penalty parameter γ is chosen not too large (about 1000) and after finding some local solutions we increase it for searching another local minimum.

The proposed algorithm starts from some initial point $x^1 \in X$. At each k-th iteration the algorithm performs randomly "drop" of two auxiliary points $\tilde{x}^1$ and $\tilde{x}^2$ and generating a curve (parabola) which passes through all three points x^k, $\tilde{x}^1$, and $\tilde{x}^2$. Then we generate some random grid along this curve and try to found all convex triples inside the grid. For each founded triple we perform refining the triple minima value with using golden section method. The best triple became an initial point for local optimization algorithm, the final point of which will be an initial point for the next iteration of global method. Details are presented in Algorithms 1 and 2.

Algorithm 1 The curvilinear multistart algorithm

Input: $x^1 \in X$—initial (start) point; $K > 0$—iterations count; $\delta > 0$; $N > 0$; $\varepsilon_\alpha > 0$—algorithm parameters.
Output: Global minimum point x^* and $f^* = f(x^*)$
1: **for** $k \leftarrow 1, K$ **do** $f^k \leftarrow f(x^k)$
2:      generate stochastic point $\tilde{x}^1 \in X$
3:      generate stochastic point $\tilde{x}^2 \in X$
4:      generate stochastic α-grid:

$$-1 = \alpha_1 \leq \ldots \leq \alpha_i \leq -\delta \leq 0 \leq \delta \leq \alpha_{i+1} \leq \ldots \leq \alpha_N = 1$$

5:      Let $\hat{x}(\alpha) = \mathrm{Proj}_X \left(\alpha^2 \left((\tilde{x}^1 + \tilde{x}^2)/2 - x^k \right) + \alpha/2 \left(\tilde{x}^2 - \tilde{x}^1 \right) + x^k \right)$ where $\mathrm{Proj}_X(z)$ - projection of point z onto set X.
     //note that $\hat{x}(-1) = \hat{x}^1, \hat{x}(1) = \hat{x}^2, \hat{x}(0) = x^k$.
6:      $f^k_* \leftarrow f^k$
7:      $\alpha^k_* \leftarrow 0$
8:      **for** $i \leftarrow 1, (N-2)$ **do**
     //Convex triplet
9:         **if** $f(\hat{x}(\alpha_i)) > f(\hat{x}(\alpha_{i+1}))$ **and** $f(\hat{x}(\alpha_{i+1})) < f(\hat{x}(\alpha_{i+2}))$ **then**
     //Refining the value of minima using
     //Golden-Section search method with accuracy ε_α
10:         $\alpha^k_* \leftarrow \mathrm{GoldenSectionSearch}(f, \alpha_i, \alpha_{i+1}, \alpha_{i+2}, \varepsilon_\alpha)$
11:         **if** $f(\hat{x}(\alpha^k_*)) < f^k_*$ **then**
12:            $f^k_* \leftarrow f(\hat{x}(\alpha^k_*))$
13:            $\alpha^k_* \leftarrow \alpha^k_*$
14:         **end if**
15:         **end if**
16:      **end for**
     //Start local optimization algorithm
17:      $x^{k+1} \leftarrow \mathrm{LOptim}(\hat{x}(\alpha^k_*))$
18: **end for**
19: $x^* \leftarrow x^k$
20: $f^* \leftarrow f(x^k)$

Algorithm 2 The local optimization algorithm

Input: $x^1 \in X$—initial (start) point; $\varepsilon_x > 0$—accuracy parameter.
Output: Local minimum point x^* and $f^* = f(x^*)$
1: **repeat**
2:      $d^k = x^k - \mathrm{Proj}_X(x^k - \nabla f(x^k))$
     //Perform local relaxation step, for example, with using standard convex interval capture technique.
3:      $x^{k+1} = \underset{\alpha \geq 0}{\mathrm{argmin}} \, f(x^k + \alpha d^k)$
4: **until** $\|x^{k+1} - x^k\|_2 \leq \varepsilon_x$

---

4 Computational Experiments

The proposed method was implemented in C language using the GNU Compiler Collection (GCC, versions: 4.8.5, 4.9.3, 5.4.0), clang (versions: 3.5.2, 3.6.2, 3.7.1,

3.8), and Intel C Compiler (ICC, version 15.0.6) on both GNU/Linux, Microsoft Windows, and Mac OS X operating systems.

We have tested the proposed algorithm on following nonzero sum six-person games. Then problem (2)–(3) is formulated as:

$$\sum_{i_1=1}^{k_1}\sum_{i_2=1}^{k_2}\cdots\sum_{i_6=1}^{k_6}\left(\sum_{q=1}^{6}a_{i_1 i_2\ldots i_6}^q\right)x_{i_1}^1 x_{i_2}^2\ldots x_{i_6}^6 - \sum_{i=1}^{6}p_i \to \max, \qquad (4)$$

$$\sum_{i_1=1}^{k_1}\sum_{i_2=1}^{k_2}\cdots\sum_{i_{q-1}=1}^{k_{q-1}}\sum_{i_{q+1}=1}^{k_{q+1}}\cdots\sum_{i_n=1}^{k_n}a_{i_1 i_2\ldots i_n}^q x_{i_1}^1 x_{i_2}^2 \ldots x_{i_{q-1}}^{q-1} x_{i_{q+1}}^{q+1}\ldots x_{i_6}^6 \le p_q \qquad (5)$$

$$i_q = 1, 2, \ldots, k_6, \quad q = 1, 2, \ldots, 6.$$

The proposed algorithm was applied for numerically solving of problem with 6 players. In all cases, Nash equilibrium points were found successfully.

Problem 3.1 We generated random 6-player game with GAMUT [15] with size $(2 \times 3 \times 3 \times 4 \times 4 \times 5)$ (totally 27 optimization variables) and the algorithm found 3 Nash equilibriums:

Points	x_q^{i*}	p_q	F^*
1	$x_1^* = (0.16, 0.84)$	67.70	5.02e–06
	$x_2^* = (0.8, 0.2, 0)$	58.30	
	$x_3^* = (0, 0, 1)$	64.30	
	$x_4^* = (0.86, 0, 0.14, 0)$	67.18	
	$x_5^* = (0, 0.98, 0, 0.02)$	73.97	
	$x_6^* = (0, 0, 0, 1, 0)$	62.71	
2	$x_1^* = (0.02, 0.98)$	48.77	−3.55e–06
	$x_2^* = (0.72, 0.28, 0)$	50.78	
	$x_3^* = (0, 0, 1)$	55.46	
	$x_4^* = (0, 0.2, 0.8, 0)$	58.61	
	$x_5^* = (0, 1, 0, 0)$	58.72	
	$x_6^* = (0, 0.58, 0.37, 0.05, 0)$	73.49	
3	$x_1^* = (0.51, 0.49)$	48.77	−1.30e–06
	$x_2^* = (1, 0, 0)$	50.78	
	$x_3^* = (0, 0, 1)$	55.46	
	$x_4^* = (0.76, 0.05, 0, 0.19)$	58.63	
	$x_5^* = (0, 0.88, 0, 0.12)$	74.14	
	$x_6^* = (0.05, 0.03, 0, 0.92, 0)$	58.86	

Problem 3.2 We generated random 6-player game with GAMUT [15] with sizes $(4 \times 5 \times 5 \times 4 \times 3 \times 2)$ (totally 29 optimization variables) and found three solutions:

Points	x^{i*}_q	p_q	F^*
1	$x^*_1 = (0.81, 0.19, 0, 0)$	55.32	$-1.02\text{e}{-}05$
	$x^*_2 = (0.24, 0, 0, 0, 0.76)$	52.30	
	$x^*_3 = (0.52, 0.14, 0, 0.05, 0.29)$	53.87	
	$x^*_4 = (0.45, 0, 0.33, 0.22,)$	53.06	
	$x^*_5 = (0, 0, 1)$	58.50	
	$x^*_6 = (0.37, 0.63)$	54.62	
2	$x^*_1 = (0, 0.84, 0.03, 0.14)$	49.36	$-2.42\text{e}{-}06$
	$x^*_2 = (0.06, 0.32, 0.35, 0.26, 0)$	51.89	
	$x^*_3 = (0.53, 0, 0, 0.47, 0)$	53.45	
	$x^*_4 = (0.31, 0.35, 0.16, 0.17)$	51.76	
	$x^*_5 = (0.32, 0.68, 0)$	51.56	
	$x^*_6 = (0.6, 0.4)$	48.28	
3	$x^*_1 = (0, 0, 0, 1)$	47.22	$-1.00\text{e}{-}04$
	$x^*_2 = (0, 0.11, 0.41, 0.14, 0.34)$	28.88	
	$x^*_3 = (0, 0.16, 0.63, 0, 0.21)$	46.00	
	$x^*_4 = (0.180.000.780.04)$	56.14	
	$x^*_5 = (0, 0, 1)$	48.13	
	$x^*_6 = (0.12, 0.88)$	45.88	

Problem 3.3 We generated random 6-player game with GAMUT [15] with sizes $(5 \times 2 \times 4 \times 3 \times 6 \times 7)$ (totally 33 optimization variables) and found three solutions:

Points	x^{i*}_q	p_q	F^*
1	$x^*_1 = (0.15, 0, 0, 0.34, 0.51)$	53.18	$5.48\text{e}{-}02$
	$x^*_2 = (0, 1)$	59.33	
	$x^*_3 = (0, 0, 0.02, 0.98)$	59.33	
	$x^*_4 = (0.210.610.17)$	38.67	
	$x^*_5 = (0, 0, 0, 0, 1, 0)$	52.60	
	$x^*_6 = (0, 0.30, 0.17, 0, 0.48, 0.05, 0)$	58.18	
2	$x^*_1 = (0, 0, 1, 0, 0)$	52.30	$-9.72\text{e}{-}05$
	$x^*_2 = (0.5, 0.5)$	36.41	
	$x^*_3 = (0.63, 0, 0.37, 0)$	51.49	
	$x^*_4 = (0.11, 0.08, 0.81)$	44.43	
	$x^*_5 = (0.37, 0.02, 0, 0, 0.14, 0.47)$	54.44	
	$x^*_6 = (0.03, 0.2, 0, 0.04, 0.18, 0.13, 0.43)$	44.86	
3	$x^*_1 = (0.06, 0.22, 0.4, 10.31, 0)$	49.62	$1.64\text{e}{-}02$
	$x^*_2 = (0.16, 0.84)$	48.69	
	$x^*_3 = (0.76, 0, 0, 0.24)$	56.47	
	$x^*_4 = (0, 0.45, 0.55)$	50.66	
	$x^*_5 = (0.35, 0.11, 0.14, 0.18, 0.22, 0)$	51.93	
	$x^*_6 = (0.36, 0.0, 80, 0.24, 0.19, 0.12, 0)$	50.38	

Problem 3.4 Game [15] with sizes ($2 \times 6 \times 6 \times 3 \times 5 \times 3$) (totally 31 optimization variables) and found solutions:

Points	x^{i*}_{q}	p_q	F^*
1	$x_1^* = (0.73, 0.27)$	43.07	−7.00e–05
	$x_2^* = (0.23, 0.23, 0.22, 0, 0.3, 20)$	49.73	
	$x_3^* = (0.21, 0.06, 0.6, 20, 0, 0.1)$	51.62	
	$x_4^* = (1, 0, 0)$	52.36	
	$x_5^* = (0, 0.23, 0, 0.77, 0)$	48.31	
	$x_6^* = (0.36, 0.28, 0.36)$	50.86	
2	$x_1^* = (0.820.18)$	51.3	−6.17e–05
	$x_2^* = (0, 0.38, 0.05, 0, 0.31, 0.26)$	51.37	
	$x_3^* = (0.26, 0, 0.43, 0.13, 0.09, 0.1)$	52.24	
	$x_4^* = (0.03, 0.63, 0.34)$	49.42	
	$x_5^* = (0.16, 0.43, 0.25, 0.1, 60)$	48.88	
	$x_6^* = (0.4, 0.31, 0.25)$	48.77	
3	$x_1^* = (0.43, 0.57)$	47.35	−1.04e–04
	$x_2^* = (0.12, 0, 0.33, 0.2, 0.36, 0)$	51.0	
	$x_3^* = (0.24, 0, 0.04, 0, 0, 0.71)$	49.83	
	$x_4^* = (0.9, 0, 0.1)$	51.52	
	$x_5^* = (0.37, 0.52, 0, 0.11, 0)$	48.16	
	$x_6^* = (0.77, 0.23, 0)$	50.32	

Problem 3.5 Game [15] with sizes ($3 \times 5 \times 5 \times 2 \times 6 \times 4$) (totally 31 optimization variables) and found solutions:

Points	x^{i*}_{q}	p_q	F^*
1	$x_1^* = (0, 0, 1)$	56.86	−1.18e–04
	$x_2^* = (0, 0, 0.41, 0, 0.59)$	81.04	
	$x_3^* = (0, 1, 0, 0, 0)$	79.20	
	$x_4^* = (1, 0)$	67.57	
	$x_5^* = (0, 0, 0, 0.82, 0, 0.17)$	58.63	
	$x_6^* = (0, 0.57, 0.28, 0.16)$	65.37	
2	$x_1^* = (0.25, 0, 0.75)$	86.12	−1.28e–04
	$x_2^* = (0.25, 0, 0, 0.75, 0)$	76.68	
	$x_3^* = (1, 0, 0, 0, 0)$	65.84	
	$x_4^* = (0, 1)$	67.11	
	$x_5^* = (0, 0, 0, 1, 0, 0)$	64.69	
	$x_6^* = (1, 0, 0, 0)$	76.99	
3	$x_1^* = (0.32, 0.59, 0.09)$	29.90	6.64e–05
	$x_2^* = (0.18, 0, 0.38, 0.4, 0.04)$	53.91	
	$x_3^* = (0.87, 0, 0, 0.13, 0)$	47.53	
	$x_4^* = (1, 0)$	38.6	
	$x_5^* = (0, 1, 0, 0, 0, 0)$	56.87	
	$x_6^* = (0, 0, 0.63, 0.37)$	45.18	

Acknowledgement This work was supported by the research grant P2018-3588 of National University of Mongolia.

References

1. Strekalovsky A.S (1998) *Global Optimality Conditions for Nonconvex Optimization*. Journal of Global Optimization 12: 415–434.
2. Enkhbat, R., Batbileg, S., Tungalag, N., Anikin, A., Gornov, A., (2017) *A Computational Method for Solving N-Person Game*. IGU Ser. Matematika ,Volume 20, 109121. https://doi.org/10.26516/1997-7670.2017.20.109
3. Enkhbat R, Tungalag N., Gornov A., Anikin A., (2016) *The Curvilinear Search Algorithm for Solving Three-Person Game*. Proc. DOOR 2016. CEUR-WS. 2016. Vol. 1623. P. 574–583. http://ceur-ws.org/Vol-1623/paperme4.pdf.
4. Strekalovsky, A. S. and Enkhbat, R. (2014) *Polymatrix games and optimization problems*. Automation and Remote Control 75(4), 632–645.
5. Orlov Andrei V. and Strekalovsky Alexander S. and Batbileg, S. (2014) *On computational search for Nash equilibrium in hexamatrix games*. Optimization Letters 10 (2): 369–381.
6. Gornov A., Zarodnyuk T. (2014) *Computing technology for estimation of convexity degree of the multiextremal function*. Machine learning and data analysis 10 (1): 1345–1353. http://jmlda.org
7. Mangasarian O. L. and Stone H. (1964) *Two-Person Nonzero Games and Quadratic Programming*. J. Mathemat. Anal. Appl. 9: 348–355.
8. Mills H. (1960) *Equilibrium Points in Finite Games*. J. Soc. Indust. Appl. Mathemat. 8 (2): 397–402.
9. Von Neumann, J., Morgenstern, O. (1944) *Theory of games and economic behavior*. Princeton University Press
10. Vorobyev N.N. (1984) *Noncooperative games*. Nauka.
11. Germeyer YU.B. (1976) *Introduction to Operation Research*. Nauka.
12. Strekalovsky A.S and Orlov,A.V. (2007) *Bimatrix Game and Bilinear Programming*. Nauka.
13. Owen,G. (1971) *Game Theory*. Saunders, Philadelphia.
14. Gibbons R. (1992) *Game Theory for Applied Economists*. Princeton University, Press.
15. Data (2018). http://www.dropbox.com/s/cbjqnzkkhph6g7y/game_6players.zip

An Infinite-Horizon Mean Field Game of Growth and Capital Accumulation: A Markov Chain Approximation Numerical Scheme and Its Challenges

Chee Kian Leong

Abstract In this paper, we characterize an infinite-horizon mean field game of growth and capital accumulation. We present a Markov chain approximation scheme (Kushner, Numerical methods for non-zero-sum stochastic differential games: convergence of the Markov chain approximation method. In: Chow PL, Yin G, Mordukhovich B (eds) Topics in stochastic analysis and nonparametric estimation. The IMA volumes in mathematics and its applications, vol 145. Springer, New York, 2008, pp 51–84) as potentially useful for obtain the numerical ϵ-Nash equilibrium solution to the infinite-horizon mean field system of equations and highlight some of the key challenges in implementing the scheme.

Keywords Mean field games · Markov-chain approximation · ϵ-Nash equilibrium

1 Introduction

Games of growth and capital accumulation have been studied extensively in economics (see, for instance, [17, 21]). Typically, such models are analyzed as infinite-horizon models, since economic growth is essentially a long-run phenomenon in economics. In this paper, we analyze an infinite-horizon mean field game of growth and capital accumulation and consider the challenges of applying a numerical scheme to solve the game.

Mean field games deal with a population of players so large that each player's action becomes asymptotically negligible. In the past, such games are modeled as static games, such as continuum population games (see [3, 23]), or as network games [2] and classical transport models [6, 7, 24]. On the other hand, the mean field game approach, pioneered by Jovanovic and Rosenthal [15] and developed further

C. K. Leong (✉)
The University of Nottingham, Ningbo, China
e-mail: chee.leong@nottingham.edu.cn

© Springer Nature Switzerland AG 2020
D. Yeung et al. (eds.), *Frontiers in Games and Dynamic Games*, Annals of the
International Society of Dynamic Games 16,
https://doi.org/10.1007/978-3-030-39789-0_8

by Huang et al. [11, 12] and Lasry and Lions [18–20], is a dynamic one solved typically within the Hamiltonian–Jacobi–Bellman framework, with the system state being described by the statistical or probability distribution of the population of individual states, namely the *mean field*, hence the term *mean field games*.

In recent years, mean field games have been applied in economics. For instance, Lucas and Moll [22] have applied mean field games to examine the role of human capital whereby individuals invest resources to improve their human capital relative to other individuals in the labor market, while Chan and Sircar [5] analyze Bertrand and Cournot games in the production of exhaustible resources. Other applications include dynamic oligopoly models [1, 25], heterogeneous agent macroeconomic models [8, 9] and auction theory [4, 14].

The mean field games of growth and capital accumulation we study in this paper were first studied by Huang and Nguyen [13]. Our current paper is distinct from Huang and Nguyen [13] in a number of aspects. We opt for an infinite-horizon economic growth model since economic growth is essentially a long-run phenomenon. Further, in their paper, Huang and Nguyen [13] make the assumption that the coefficient of the relative risk aversion is equal to the capital share, which vastly simplifies the mean field system of equations and its solution. While mathematically and analytically useful, such assumption may not be economically tenable since there is no a priori reason that the coefficient of risk aversion is equal to the capital share. In our paper, we do not resort to this assumption. This results in a mean field system of equations based on the original set of parameters of the model. However, solving an infinite-horizon mean field games can be particularly challenging, even when a numerical scheme is involved, since unlike the finite-horizon games considered in Huang and Nguyen [13], there is no boundary condition for the infinite-horizon mean field games. We propose a numerical scheme based on the Markov chain approximation scheme [16] to derive the ϵ-Nash equilibrium in the mean field game. Yet the implementation of this scheme can be challenging. Some of these challenges are discussed in this paper.

This paper is organized as follows. Section 2 presents the infinite- horizon mean field game of growth and capital accumulation. Section 3 outlines its solution. Section 4 presents a potential numerical scheme to solve the infinite-horizon mean field game and also discusses some of the difficulties. Concluding remarks are presented in Sect. 5.

2   Mean Field Games of Growth and Capital Accumulation

Consider an economy with households and firms. The economy has an aggregate production function, with the aggregate output (Y) being produced according to technology

$$Y_t = f(K_t, L_t)$$

where the aggregate capital $K_t = \sum_{i=1}^{n} k_t^i$ and aggregate labor supply $L_t = \sum_{i=1}^{n} l_t^i$. In intensive form, this may be written as $y_t = f(k_t)$, where $y_t = Y_t/L_t$ and $k_t = K_t/L_t$.

There is a finite population of n households, which are heterogeneous in their capital stock, denoted by k^i for the ith household. Each household i determines its optimal consumption by solving the intertemporal utility maximization problem with a constant discount rate ρ

$$\max_{c_t^i} E \int_0^\infty e^{-\rho t} u\left(c_t^i, c_t^E\right) dt$$

where c_t^i is consumption of the ith household and c_t^E is the expected consumption in the population. Hence, each household compares its consumption relative to the expected consumption in the population and thereby derives utility.

Household i owns capital k^i, which is provided to the firms in the economy and earns a competitive interest rate r, which is determined by the aggregate capital in the economy, that is, K. At the same time, it supplies labor l_t^i competitively at the wage rate w, which again depends on the aggregate labor supplied in the economy, that is, L_t. Denote k_0^i are the given initial capital stock for household i. Capital is accumulated via the Ito stochastic differential equation (SDE)

$$dk^i = [w + rk^i - c^i]dt + \sigma k^i dW^i$$

where dW^i is the increment of a standard Wiener (white noise) or Brownian motion process W^i, independently and identically distributed across households, while σk^i can be interpreted as the uncertainty associated with the capital stock.

All firms are identical, so we assume a representative firm which solves the (aggregate) profit maximization problem

$$\max_{K,L} Af(K, L) - (r + \delta)K - wL$$

where A is fixed productivity (for simplicity, we assume $A = 1$), K is the capital, L labor supply, $f(\cdot)$ is an increasing and concave production function satisfying the Inada condition, δ depreciation rate and $r + \delta$ can be interpreted as the user cost of capital. This problem can be simplified by recasting the optimization problem in intensive form and normalizing L to 1. In equilibrium, we can characterize $\bar{k}$ as the expected value of k_t^i, given some probability density function $g\left(k_t^i\right)$:

$$\bar{k} = \int k^i g\left(k^i\right) dk^i$$

It is possible to specify this in terms of higher moments of $g\left(k^i\right)$, but for simplicity, we will only consider the first moment of this distribution.

Since t does not appear in the objective function and the state dynamics, we are dealing with an infinite-horizon autonomous mean field game.

Next, we characterize the equilibrium wage and capital rent in terms of this expected value as follows:

$$\widetilde{w} = f(\overline{k}) - kf'(\overline{k}),$$

$$\widetilde{r} = f'(\overline{k}) - \delta.$$

Using these equilibrium values, we obtain the household's Hamilton–Jacobi–Bellman equation (HJB)

$$\rho V(k) - \frac{\sigma^2}{2} V_{kk}(k) = \max_c \left\{ u(c) + V_k(k) \left[\widetilde{w} + \widetilde{r}k - c \right] \right\}. \tag{1}$$

The distribution of capital is described by the Fokker–Planck–Kolmogorov (FPK) forward equation

$$\frac{\partial g(k)}{\partial t} = -\frac{\partial}{\partial k} \left[\mu^k(k) g(k) \right] + \frac{\sigma^2}{2} \frac{\partial^2}{\partial k^2} g(k), \tag{2}$$

where

$$\mu^k(k) = \widetilde{w} + \widetilde{r}k - \widetilde{c}(k)$$

and

$$\widetilde{c}(k) = \varphi \left(\overline{k} | \mu^{\overline{k}} \right) \tag{3}$$

is the optimal consumption choice derived from the HJB. The feedback control $\varphi\left(k | \mu^k\right)$ can be interpreted as the best response for any individual household with respect to the infinite population, which is represented by the measure μ^k. Together, these equations form the mean field system of equations for the infinite-horizon mean field game of growth and capital accumulation.

3   Solution

There are a variety of solution concepts to mean field games. Huang and Nguyen [13] employ the Nash certainty equivalence principle [11, 12]. For discrete-time population dynamic games with Markov decision process, Weintraub et al. [25] also obtain the closely related concept of oblivious equilibria. Like differential games, the characterization of the information in a mean field game is important. There are various ways to characterize information: (1) an open loop control will be solely a function of the information set containing the time and the initial state of that agent or of the global initial state; (2) a Markovian (closed loop or feedback) control will be a function of current time and the current state. Open-loop strategies are necessarily static in nature; however, in mean field games with a large number of small players, open-loop strategies could be useful as an approximation. The different specifications of the information structure lead to different solution methods and concepts in the mean field game literature.

Another issue is the information available to an individual agent about other agent's dynamics. Typically, an agent has information on its own dynamics (more specifically, the structure and parameters of the dynamic optimization problem), but the uncertainty about other agent's dynamics is captured in the form of a probability distribution over the population of agents. Thus the individual agent is not able to identify precisely the dynamics of another specific individual agent. In this case, the agents will determine their best response in reaction to the best responses of all other agents based on the system dynamics. Consequently, while the agent can determine the population control strategy, it cannot determine the specific control action of any other particular agent, since the state distribution of the latter is not available to the first agent.

For our infinite-horizon mean field game of growth and capital accumulation, we consider the ϵ-Nash equilibria, which can be characterized as follows.

Consider a set of households h_i, $1 \leq i \leq n$, let $C^n = C_1 \times C_2 \times \ldots \times C_n$ denote the joint admissible consumption space, where each space consists of a set of consumption strategies c_i, $1 \leq i \leq n$. A joint consumption strategy is denoted by $c = (c_1, c_2, \ldots, c_n)$.

Definition The joint strategy $c^* = \left(c_1^*, c_2^*, \ldots, c_n^*\right)$ is an ϵ-Nash equilibrium, for some small $\epsilon \geq 0$, if

$$\sup_{c_i} J_i \left(c_i, c_{-i}^*\right) \leq J_i \left(c_i^*, c_{-i}^*\right) \leq J_i \left(c_i^*, c_{-i}^*\right) + \epsilon$$

.

Thus, we are saying that when all households except for household i conforms to the optimal consumption strategies c_{-i}^*, the unilateral deviation of household i will yield a gain of at most ϵ. When $\epsilon = 0$, we obtain back our closed-loop Nash equilibrium.

In practice, the approach to solving the mean field game involves specifying some functional forms for $u(c_t)$ and $F(K_t, L_t)$. We can specify a CRRA utility and a Cobb–Douglas production function as follows:

$$u\left(c_t, c_t^E\right) = \frac{1}{1-\gamma}\left(\frac{c_t}{c_t^E}\right)^{1-\gamma}, \quad \gamma > 0$$

$$f(k_t) = k_t^\alpha, \quad \alpha > 0$$

with $c_t^E > 0$.

As such, we obtain

$$\widetilde{w} = (1-\alpha)\,\overline{k}^{\alpha},$$

$$\widetilde{r} = \alpha\overline{k}^{\alpha-1} - \delta,$$

where $\overline{k}$ is obtained via the FPK forward equation for the distribution of capital

$$\frac{\partial g\,(\overline{k})}{\partial t} = -\frac{\partial}{\partial k}\left[\mu\,(\overline{k})\,g\,(\overline{k})\right] + \frac{\sigma^2}{2}\frac{\partial^2}{\partial\overline{k}^2}g\,(\overline{k}) \qquad (4)$$

for some probability distribution $g\,(\overline{k})$ where

$$\mu\,(\overline{k}) = (1-\alpha)\,\overline{k}^{\alpha} + \widetilde{r}_t\alpha\overline{k}^{\alpha-1} - \widetilde{r}_t\delta - \widetilde{c}\,(k)$$

and

$$\widetilde{c}\,(k) = \varphi\left(\overline{k}|\mu^{\overline{k}}\right). \qquad (5)$$

To obtain the optimal $\widetilde{c}\,(k)$, we can solve the HBJ equation in two stages, similar to that adopted by Huang and Nguyen [13]. In the first stage, we assume that c_t^E is fixed as c^E and solve for the optimal $\hat{c}^i$ given c^E. In the second stage, we use the optimal $\hat{c}$ to write the state equation

$$dk^i = [\widetilde{w} + \widetilde{r}k^i - \hat{c}^i]dt + \sigma k^i dW^i$$

while imposing the consistency condition $\overline{c} = \sum_{i=1}^{N}\hat{c}^i$.

If $V_k\,(k,t) > 0$ holds, then the maximum value of

$$\max_c\left\{\frac{1}{1-\gamma}\left(\frac{c}{c^E}\right)^{1-\gamma} + V_k\,(k,t)\left[(1-\alpha)\,k_t^{\alpha} + \left(\alpha k^{\alpha-1} - \delta\right)^2 - c\right]\right\}$$

will be given by

$$\hat{c} = \left[\left(c^E\right)^{1-\gamma}V_k\,(k).\right]^{-\frac{1}{\gamma}}$$

We use this value to rewrite the HBJ as follows:

$$\rho V\,(k) - \frac{\sigma^2}{2}V_{kk}\,(k) = \frac{\gamma}{1-\gamma}\left(c^E\right)^{\frac{\gamma-1}{\gamma}}\left[V_k\,(k)\right]^{\frac{\gamma-1}{\gamma}}$$

$$+ V_k\,(k)\left[(1-\alpha)\,k^{\alpha} + \left(\alpha k^{\alpha-1} - \delta\right)^2\right],$$

and the capital accumulation SDE:

$$dk = \left\{ w + rk - \left[\left(c^E \right)^{1-\gamma} V_k\left(k\right) \right]^{-\frac{1}{\gamma}} \right\} dt + \sigma k dW.$$

Huang and Nguyen [13] employ an ansatz with boundary conditions to obtain their solution to their finite horizon problem. However, this is not possible for infinite horizon problems because of the lack of boundary conditions. Moreover, the ansatz approach introduces two new variable functions, both of which the economic intuitions may not be so evident.

4   Markov Chain Approximation Scheme

In this section, we consider a numerical approach based on the Markov chain approximation scheme [16] to derive the ϵ-Nash equilibrium in the mean field game.

In a nutshell, this Markov chain approximation scheme involves three parts:

1. We replace the state space with a finite grid.
2. The partial derivatives of V are substituted with their finite differences.
3. We approximate the stochastic state trajectory using a Markov chain whose transition probabilities are derived from the characteristics of the state process.

First, define R_h^n as the h-grid on R^n:

$$R_h^n = \left\{ k : k = \sum_{i=1}^{n} z_i e_i h \right\}$$

for some integer z_i and unit vector e_i in the ith direction.

We use the following rules for approximating the partial derivatives

$$V_k\left(k\right) = \begin{cases} \frac{V(k+e_ih)-V(k)}{h} & \text{if } \widetilde{w} + \widetilde{r}k - \hat{c} \geq 0 \\ \frac{V(k)-V(k-e_ih)}{h} & \text{if } \widetilde{w} + \widetilde{r}k - \hat{c} < 0 \end{cases}$$

$$V_{kk}\left(k\right) = \frac{V\left(k+e_ih\right) - V\left(k-e_ih\right) - 2V\left(k\right)}{h^2}.$$

Define the interpolation interval

$$\Delta t^h\left(k, \mathbf{c}\right) = \frac{h^2}{Q_h\left(k, \mathbf{c}\right)},$$

where

$$Q_h(k, \mathbf{c}) = \sigma^2 + h \sum_{i=1}^{n} |\tilde{w} + \tilde{r}k - \hat{c}|.$$

Next, define the transition probability functions in the Markov chain

$$\pi^h(k, k \pm e_i h) = \frac{\frac{\sigma^2}{2} + \tilde{w} + \tilde{r}k - \hat{c}}{Q_h(k, \mathbf{c})}$$

$$\pi^h(k, k + e_i h + e_j h) = \pi^h(k, k - e_i h - e_j h)$$

$$= 0$$

$$\pi^h(k, k + e_i h - e_j h) = \pi^h(k, k - e_i h + e_j h)$$

$$= 0$$

$$\pi^h(k, k) = \frac{h^2}{Q_h(k, \mathbf{c})}.$$

Substitute and rearrange terms to obtain the following equation

$$\rho V(k) - \frac{\sigma^2}{2} \left[\frac{V(k + e_i h) - V(k - e_i h) - 2V(k)}{h^2} \right]$$

$$= \beta [V_k(k)]^{\frac{1}{\beta}} + V_k(k) \left[(1 - \alpha) k^\alpha + \left(\alpha k^{\alpha-1} - \delta \right)^2 \right]$$

by setting $\beta = \frac{\gamma}{1-\gamma}$ and using the appropriate $V_k(k)$. For instance, if $\tilde{w} + \tilde{r}k^i - \hat{c}^i \geq 0$, we have

$$\left[\rho + \frac{\sigma^2}{h^2} + \frac{g(k)}{h} \right] V(k) = \frac{\sigma^2}{2} \left[\frac{V(k + e_i h) - V(k - e_i h)}{h^2} \right]$$

$$+ \beta \left(c^E \frac{V(k + e_i h) - V(k)}{h} \right)^{\frac{1}{\beta}} + \frac{g(k)}{h} V(k + e_i h).$$

This equation can be solved as the fixed point problem of a discrete-state dynamic programming operator

$$V(k) = \Gamma(k)$$

with the operator Γ contracting.

The Markov chain obtained from the scheme described earlier will provide a converging approximation of the value function for the player i [16]. To establish the fixed point, Howard's policy improvement algorithm [10] may be employed.

Finally, the c^{E*} obtained using this numerical scheme could be substituted back to obtain $\widetilde{c}(k) = \left[\left(c^{E*} \right)^{1-\gamma} V_k(k) \right]^{-\frac{1}{\gamma}}$, with which the FPK equation can be solved to obtain the distribution of the capital in the economy using some appropriate capital distribution function $g(k)$.

Despite its attractiveness, implementing this numerical scheme is not without its difficulties. The first is a grid dimensionality issue. The scheme is particularly sensitive to the fineness of the grid and may not easily converge to a solution. Another issue is the appropriate initial choice of c^E. The choice is currently arbitrary and the numerical scheme is not yet robust to different choices of c^E. This can be particularly problematic, as in some simulation runs, the numerical scheme converges to the initial value of c^E. Once the c^{E*} is obtained, solving the FPK requires another assumption about the capital distribution function $g(k)$. Deciding on the appropriate capital distribution function is important since different choices of $g(k)$ will result in different economic interpretations.

5 Concluding Remarks

In this paper, we consider an infinite-horizon mean field game of growth and capital accumulation. Finding the ϵ-Nash equilibrium solutions to the infinite-horizon mean field games can be particularly challenging, even when a numerical scheme is involved, since unlike the finite-horizon games considered in Huang and Nguyen [13], there is no boundary condition for the infinite-horizon mean field games. Ideally, the mean field system of equations should be solved together and updated accordingly. In practice, the numerical scheme generally starts with the solution to the HBJ problem, followed by the FPK problem. In this paper, we present one possible numerical scheme based on the Markov chain approximation scheme [16] to derive the ϵ-Nash equilibrium in the mean field game. However, the implementation of this scheme involves some challenges, such as grid dimensionality issue and appropriate choices of initial values and the capital distribution functions. Consequently, finding an appropriate and robust numerical scheme to solve infinite-horizon mean field games remains a challenging research question in the mean field games literature.

References

1. Adlakha, S, Johari, R., Weintraub, G. Y. (2015) Equilibria of dynamic games with many players: existence, approximation and market structure. Journal of Economic Theory, 156: 269–316.
2. Altman, E., Basar, T., Srikant, R. (2012) Nash equilibria for combined flow control and routing in networks: asymptotic behavior for a large number of users. IEE Transactions of Automatic Control, 47(6): 917–930.

3. Aumann, R.J. and Shapley, L.S. (1974) Values of non-atomic games. Princeton, NJ: Princeton University Press.
4. Balseiro, S.R., Besbes, O., Weintraub, G.Y. (2015) Repeated auctions with budgets in ad exchanges: Approximations and design. Management Science, 61(4): 864–884.
5. Chan, P. and Sircar, R. (2015) Bertrand and Cournot mean field games. Applied Mathematics & Optimization, 71(3), 533–569.
6. Correa, J.R, Stier-Moses, N.E. (2010) Wardrop equilibria. In: Cochran, J.J. (ed) Wiley Encyclopedia of Operation Research and Management Science. Hoboken, US: John Wiley & Sons.
7. Haurie, A., Macrotte, P. (1985) On the relationship between Nash-Cournot and Wardrop equilibria. Networks, 15(3): 295–308.
8. Heathcote, J., Storesletten, K., Violante, G.L. (2009) Quantitative macroeconomics with heterogeneous households, Annual Reviews in Economics, 1: 319–354.
9. Hopenhayn, H.A. (1992) Entry, exist and firm dynamics in long run equilibrium, Econometrica, 60(5) 1127–1150.
10. Howard, R. (1960) Dynamic Programming and Markov Processes. Mass., CA: MIT Press.
11. Huang, M., Caines, P.E., Malharme, R.P. (2003) Individual and mass behavior in large population stochastic wireless power control problems: centralized and Nash equilibria solutions. In: Proceedings of the IEEE Conference on Decision and Control, Maui. pp 98–103.
12. Huang, M., Caines, P.E., Malharme, R.P. (2007) Large-population cost-coupled LQG problems with non-uniform agents: individual-mass behavior and decentralized ϵ-Nash equilibria. IEEE Transactions on Automatic Control, 51: 1560–1571.
13. Huang, M., Nguyen, S.L. (2016) Mean field games for stochastic growth with relative utility. Applied Mathematics & Optimization, 74: 643–668.
14. Iyer, K., Johari, R., Sundararajan, M. (2014) Mean field equilibria of dynamic auctions with learning, Management Science, 60(12), 2949–2970.
15. Jovanovic, B., Rosenthal, R.W. (1988) Anonymous sequential games. Journal of Mathematical Economics, 17(1): 77–87.
16. Kushner H.J. (2008) Numerical methods for non-zero-sum stochastic differential games: Convergence of the Markov chain approximation method. In: Chow P.L., Yin, G., Mordukhovich, B. (eds) Topics in Stochastic Analysis and Nonparametric Estimation. The IMA Volumes in Mathematics and its Applications, vol 145.New York, NY: Springer, pp 51–84.
17. Lancaster, K. (1973) The dynamic inefficiency of capitalism. Journal of Political Economy, 81, 1092–1109
18. Lasry J.M., Lions P.L. (2006a) Jeux a champ moyen. I - Le cas stationnaire. Comptes Rendus Mathematique, 343(9):619–625.
19. Lasry, J.M. and Lions, P.L. (2006b) Jeux a champ moyen. II - Horizon fini et controle optimal. Comptes Rendus Mathematique, 343(10):679–684.
20. Lasry, J.M. and Lions, P.L. (2007) Mean field games, Japanese Journal of Mathematics, 2(1): 229–260.
21. Leong, C.K. and W. Huang (2010) A stochastic differential game of capitalism. Journal of Mathematical Economics, 46 (4): 552–561.
22. Lucas, R.E. and Moll, B. (2014) Knowledge growth and the allocation of time. Journal of Political Economy, 122(1): 1–51.
23. Neyman, A. (2002) Values of games with infinitely many players. In: Aumann, R. J., Hart, S. (eds) Handbook of Game Theory, Vol 3. North Holland, Amsterdam, pp 2121–2167.
24. Wardrop, J.G. (1952) Some theoretical aspects of road traffic research. In: Proceedings of the Institute of Civil Engineers, London, Part II, Vol 1, pp 325–378.
25. Weintraub, G.Y, Benkard, C.L. and Van Roy, B. (2008) Markov perfect industry dynamics with many firms, Econometrica, 76(6): 1379–1411.